AF411357

Characterization of Bioactive Compounds in Foods and Plants Using Advanced Analytical Techniques

Characterization of Bioactive Compounds in Foods and Plants Using Advanced Analytical Techniques

Editors

Marina Russo
Francesco Cacciola

MDPI • Basel • Beijing • Wuhan • Barcelona • Belgrade • Manchester • Tokyo • Cluj • Tianjin

Editors
Marina Russo
Chibiofaram
University of Messina
Messina
Italy

Francesco Cacciola
Biomorf
University of Messina
Messina
Italy

Editorial Office
MDPI
St. Alban-Anlage 66
4052 Basel, Switzerland

This is a reprint of articles from the Special Issue published online in the open access journal *Foods* (ISSN 2304-8158) (available at: www.mdpi.com/journal/foods/special_issues/characterization_bioactive_analytical_techniques).

For citation purposes, cite each article independently as indicated on the article page online and as indicated below:

LastName, A.A.; LastName, B.B.; LastName, C.C. Article Title. *Journal Name* **Year**, *Volume Number*, Page Range.

ISBN 978-3-0365-2855-7 (Hbk)
ISBN 978-3-0365-2854-0 (PDF)

Contents

Preface to "Characterization of Bioactive Compounds in Foods and Plants Using Advanced Analytical Techniques"

Dear Colleagues,

Since the 1990s, food chemistry opened a new chapter in foods and plants investigation. An increasing attention to secondary metabolites and micro-constituents of nutraceutical interest present in foods has been noticed, supporting previous studies on macronutrient composition. Thanks to positive scientific opinions on the presence of bioactive molecules in plants and foods, the previous vision of exploring foods exclusively from a "caloric" point of view has been changed to looking at foodstuffs as having positive effects on human health.

This book focuses on the optimization and validation of advanced analytical methodologies dedicated to the characterization and valorization of foods and plants containing bioactive molecules. Qualitative and quantitative characterization, food security, traceability, and innovation in the field of nutraceutical and functional nutrition will be of particular interest in order to stimulate a dialogue on correct nutrition concepts in a constantly changing cultural, technological, and climate context.

Marina Russo, Francesco Cacciola
Editors

 *foods*

Article

Characterization of Phytochemicals in Berry Fruit Wines Analyzed by Liquid Chromatography Coupled to Photodiode-Array Detection and Electrospray Ionization/Ion Trap Mass Spectrometry (LC-DAD-ESI-MS^n) and Their Antioxidant and Antimicrobial Activity

Agata Czyżowska *[iD], Agnieszka Wilkowska[iD], Agnieszka Staszczak (Mianowska) and Agnieszka Nowak[iD]

Institute of Fermentation Technology and Microbiology, Faculty of Biotechnology and Food Sciences, Lodz University of Technology, 171/173 Wolczanska Street, 90-924 Lodz, Poland; agnieszka.wilkowska@p.lodz.pl (A.W.); agnieszkamianowska@op.pl (A.S.); agnieszka.nowak@p.lodz.pl (A.N.)
* Correspondence: agata.czyzowska@p.lodz.pl

Received: 29 September 2020; Accepted: 26 November 2020; Published: 1 December 2020

Abstract: Fruits are a valuable source of phytochemicals. However, there is little detailed information about the compounds contained in fruit wines. In this study, wines from six different berries were analyzed using HPLC-DAD-ESI-MS^n. About 150 compounds were identified, including anthocyanins (34), hydroxycinnamic acids (12) and flavonols (36). Some of the compounds were identified for the first time in berry wines. The blackberry wines were found to contain the largest number of bioactive compounds (59). Elderberry wines where the richest source of polyphenols (over 1000 mg/L) and contained the largest amounts of all of the analyzed groups of compounds (hydroxycinnamic acids, anthocyanins and flavonols). The lowest concentration of polyphenols was observed in the wines made from cranberries and bilberries (less than 500 mg/L). The antioxidant activity was determined in relation to ABTS$^+$, DPPH, and FRAP. The highest values were observed in the blackberry wines, and the lowest for the cranberry wines. The wines were analyzed to test their antimicrobial activity. Five of the six wines (with the exception of elderberry wine) inhibited *Bacillus cereus* growth and two (blackberry and cranberry wines) were active against *Listeria monocytogenes*.

Keywords: berries; fruit wines polyphenols identification; LC–MS^n

1. Introduction

Fruits are known to be a valuable source of phytochemicals. However, there is little detailed information in the literature about the compounds contained in fruit wines [1–9]. Only fruit wines from strawberries and bilberries (*Vaccinium myrtillus* L.) have been studied intensively [10–12]. Behrend and Weber [10] analyzed the anthocyanins and tannins in bilberry wines fermented after different pretreatments and during ageing. Liquid chromatography–mass spectrometry (LC–MS) of the anthocyanins was performed using an LTQ-XL ion trap mass spectrometer connected to a UHPLC system via an electrospray ionization (ESI) interface. The anthocyanin profiles of all the wines were identical at the start of fermentation. During fermentation, considerable changes were noticed. The wines that had been subjected to prefermentative thermal treatment had an almost juice-like composition. The other wines displayed lower levels of arabinosides and galactosides. Polymeric pigments and pyranoanthocyanins were observed in all the wines. Liu et al. [12] analyzed bilberry wines by liquid chromatography using a diode array detector and electrospray ionization-quadrapole/time of flight hybrid mass spectrometry (ESI-QTOF-MS). They identified 42 nonanthocyanin compounds, including 22 phenolic acids, 15 flavonols and 5 flavan-3-ols. Hornedo-Ortega et al. [11] analyzed the anthocyanins in strawberry beverages. The anthocyanin fraction of the fermented strawberry wine was analyzed on an Amberlite XAD7HP column. Four anthocyanin compounds were identified with high accuracy for the first time in strawberry wines: pelargonidin-3-sambubioside, pelargonidin disaccharide (hexose + pentose) acylated with acetic acid, cyanidin-3-(6-acetyl)-glucoside, and pelargonidin 3-(6-succinyl)-arabinoside/3-(6-malonyl)-rhamnoside.

Papadopoulou et al. [13] demonstrated the antibacterial activity of the polyphenols in various white and red wines against strains of *Staphylococcus aureus* and *Escherichia coli*. Fruit wines, which are also a rich source of polyphenolic compounds and other bioactive compounds [2–9], may have similar antibacterial properties.

Polyphenols can also show antifungal activity, although it is much weaker [13,14]. Moreover, via various mechanisms, polyphenolic compounds limit the acquisition of resistance by microorganisms [15].

Red and purple fruits are rich sources of anthocyanins, which show bacteriostatic and bactericidal activity against many microorganisms (including *Staphylococcus* sp., *Klebsiella* sp., *Helicobacter* and *Bacillus*) [16,17]. Raspberries and cloudberries are rich sources of ellagitannins. These compounds are also found in strawberries, but in smaller amounts [17,18]. Quercetin is another compound found in fruits. It has been shown to increase the permeability of bacterial cell wall, which may, for example, increase the sensitivity of bacteria to antibiotics [19]. In many cases, mixtures of these compounds have been found to have stronger effects than any of their components separately. Of the various berries, raspberries and cloudberries (*Rubus chamaemorus*) are recognized as the best inhibitors of bacteria such as *Staphylococcus* spp., *Salmonella* spp., *Helicobacter pylori* and *Bacillus cereus* [20,21]. Raspberry juice has been reported to completely inhibit the growth of *Escherichia coli* in vitro [22].

Berries are also a rich source of vitamins A, C, and E. Ascorbic acid is found in a wide variety of fresh fruits [23]. In addition to having redox potential, it is also an excellent electron donor in biological systems [24]. Epidemiological and experimental evidence suggests that vitamin C can protect against the development of gastric cancer by several potential mechanisms: it reduces gastric mucosal oxidative stress, DNA damage, and gastric inflammation by scavenging ROS (reactive oxygen species); it inhibits gastric nitrosation and the formation of N-nitroso compounds by reducing nitrous acid to nitric oxide and producing dehydroascorbic acid in the stomach; it enhances host immunologic functions; it has a direct effect on *Helicobacter pylori* growth and virulence; it inhibits gastric cell proliferation and induces apoptosis [25]. The content of vitamin C in berry fruits can be influenced by numerous factors, including the species, variety, weather conditions, ripeness, and region [23]. Vitamin stability can be affected by various technological practices used during the processing of food, namely changes in temperature (e.g., thermal treatments) and oxygen levels [24]. The concentration of vitamins decreases during winemaking (fermentation and ageing) [26].

In this study, we characterize and quantify the bioactive compounds in wines made from six different berries, using HPLC-DAD-ESI-MS^n. We also investigate whether the fruit wines may be considered a source of antimicrobial agents against pathogenic microorganisms. The fruit wines were tested against both pathogenic Gram-negative bacteria (*E. coli* and *Salmonella* Enteritidis), which can cause foodborne and waterborne outbreaks of gastrointestinal tract infections, and pathogenic Gram-positive bacteria (*B. cereus*, *L. monocytogenes* and *S. aureus*), which can cause food poisoning and toxic symptoms in humans. *Candida albicans* ATCC 10231, was used as a reference strain for the analysis of antifungal action [27]. Some of these microbes can also colonize oral human cavities [28–32]. Studies suggest that between 94% and 100% of healthy adults have oral colonization with *Staphylococcus* spp. [30] and oral carriage of *S. aureus* ranges from 24% to 36% [31].

2. Materials and Methods

2.1. Reagents and Standards

ABTS$^{+\bullet}$ (2,2′-azinobis 3-ethylbenzothiazoline-6-sulfonic acid), potassium persulphate, FeCl$_3$, TPTZ (2,4,6-Tris (2-pyridyl-S-triazine), DPPH$^{+\bullet}$ (2,2-diphenyl-1-picrylhydrazyl) and methanol were purchased from Sigma (Poznań, Poland). Formic acid and HPLC-grade acetonitrile were sourced from J.T. Baker (Witko, Poland). Anthocyanin standards were produced by Extrasynthese (Genay, France) and PhytoLab (Vestenbergsgreuth, Germany). Available standards of other polyphenols were purchased from Sigma (Poznań, Poland) and Extrasynthese (Genay, France). HPLC-grade water was obtained using an Aquinity E60 Lifescience TI system (membraPure GmbH, Bodenheim, Germany).

2.2. Wine Preparation

Fruit wines were prepared according to the Polish Law of 12 May 2011 'On the production and bottling of wine products, trade and organization of the wine market'. Six wine types were made from the following berries: bilberry (common bilberry) (*Vaccinium myrtillus* L.)—BB; blackberry (*Rubus* L.)—B; cranberry (*Vaccinium macrocarpon* Aiton)—C; elderberry (*Sambucus nigra* L.)—E; raspberry (*Rubus idaeus* L.)—R; and strawberry (*Fragaria × ananassa*)—S (Kent variety).

Fresh blackberry, cranberry, raspberry and strawberry fruits (about 10 kg of each species) were purchased from local retailers between June and October, depending on the availability. Bilberry fruits were collected in the region of Belchatow (51°21′ N, 19°21′ E) and elderberry fruits in Pabianice (51°39′ N, 19°21′ E). The elderberry stalks were removed. The fruits were then heat treated (85 °C, 5 min) to inactivate polyphenol oxidase-type enzymes. The blueberry, elderberry and cranberry pulps were cooled to 50 °C and treated with pectinolytic enzyme (Rohapect 10 L, AB Enzymes GmbH, Darmstadt Germany, AKE, Pabianice, Poland) at a dose of 0.5 g/kg of fruits. They were then pressed using a hydraulic press.

Fermentation was performed at 25 °C using BCS103 wine yeast (Fermentis, LeMag, Żyrardów, Poland) at a dose of 0.2 g/L. Once fermentation was complete, the wine was racked and poured into bottles. All the wines were aged for around 5 months. The wines were then subjected to basic analysis (alcohol, extract, sugar, acidity). The wines were dealcoholated and tested for their antimicrobial activity.

2.3. Preparation of Dealcoholated Red Wines (DRW)

To remove the alcohol from the wines, an equal volume of distilled water was added to a given volume of wine and then concentrated to the original volume (38 mbar, 35 °C, 140–180 rev/min). The solutions were concentrated on a Büchi vacuum evaporator—Rotavapor R-215 (Büchi Labortechnik AG, Flawil, Switzerland).

2.4. Analysis of Organic Acids, Sugars and Alcohols

Organic acids, glucose, fructose and alcohols (ethanol and glycerol) were analyzed using a Finnigan Surveyor HPLC system (Thermo Fisher Scientific Inc., Waltham, MA, USA), according to the method described by Czyżowska et al. [33].

2.5. Total Phenolic Content (TPC) Assay

Total phenolic content (TPC) was determined using the Folin–Ciocalteu reaction with gallic acid as a standard. To a test tube were added 0.1 mL of the 5-fold diluted sample, 0.2 mL of Folin–Ciocalteu reagent, 1 mL of 20% sodium carbonate and 2 mL of distilled water. In the control, 0.1 mL of distilled water was added instead of the test solution. The samples were mixed and incubated for 1 h at room temperature, in the absence of light. After incubation, the absorbance was measured at a wavelength of λ = 765 nm (Cecil CE 2041, Cecil Instruments Limited, Cambridge, UK).

2.6. Analysis of Antioxidant Capacity

2.6.1. ABTS Radical-Scavenging System

Radical scavenging activity against $ABTS^{+\bullet}$ was determined based on the method described by Rivero-Pérez et al. [34] with slight modifications. The wine solution (0.2 mL) was mixed with 4 mL of ABTS reagent and incubated for 15 min. The results were expressed as mM of Trolox equivalents, using linear calibration obtained with different concentrations of Trolox.

2.6.2. DPPH Radical-Scavenging System

The method described by Fogliano et al. [35] was applied with slight modifications. Wine solution (0.2 mL) was mixed with 4 mL of $DPPH^{+\bullet}$ reagent (65 μM) and incubated for 30 min. Absorbance was measured at 515 nm. The results were expressed as mM of Trolox equivalents on the relevant calibration curve.

2.6.3. FRAP Method

The method described by Rivero-Pérez et al. [34] was used with slight modifications. For 30 min, 2.9 mL of the reactive mixture was incubated with 50 μL of the sample. Absorbance was measured at 595 nm. The results were expressed as mM of Trolox equivalents on the relevant calibration curve.

2.7. LC–MS^n Identification of Wines Compounds

Qualitative analysis of the bioactive compounds in the berry wines was conducted using an HPLC coupled on-line with an MS LTQ Velos mass spectrometer (ThermoScientific, Waltham, MA, USA), following the method described by Efenberger-Szmechtyk et al. [36]. Separation was carried out using a Hypersil Gold column (150 × 2.1, particle size 1.9 μm) (ThermoScientific, Waltham, MA, USA). The column was thermostated at 45 °C. For the anthocyanins, 2.5% formic acid solution (phase A) and 95% acetonitrile (phase B) were used as eluents with a flow rate of 220 μL/min and an injection volume of 10 μL. For other compounds, the mobile phase consisted of solvent A (1 mL formic acid in 1 L of deionized water) and solvent B (95% acetonitryl). The separation was carried out with the following gradients: in the first 8 min, a linear gradient from 96% to 85% phase A; 8–12 min linear gradient from 85% to 82% phase A; 12–40 min linear gradient from 82% to 60% phase A; 40–44 min linear gradient from 60% to 50% phase A; 44–47 min linear gradient from 60% to 50% phase A; 47–49 min linear gradient from 50% to 96% phase A, followed by column recalibration.

Spectrometry was performed with a capillary voltage of 4 kV and collision energy of 20 V. The desolvation temperature was 280 °C and the source temperature was 350 °C.

Detection of anthocyanins was carried out in positive ion mode, whereas the other compounds were detected in the negative ion mode in the range of m/z from 100 to 1200. The compounds were identified based on a comparison of the maximum absorption spectra of UV radiation. The molecular weight was determined on the basis of the mass to charge ratio. Retention times and fragmentation spectra were compared with the available standards and literature data.

2.8. HPLC Analysis of Polyphenols

Prior to analysis, the samples were filtered through a 0.45 mm membrane and injected into the HPLC system. HPLC-PDA analyses were performed using a Finnigan Surveyor equipped with an autosampler, a diode array detector Finnigan Surveyor-PDA Plus (Thermo FisherScientific Inc., Waltham, MA, USA) and ChromQuest 5.0 chromatography software (Thermo FisherScientific Inc., Waltham, MA, USA). The separation conditions were as described by Efenberger-Szmechtyk et al. [36]. The calibration curves were established using standards for chlorogenic acid, quercetin-glucoside and cyanidin-glucoside to quantify polyphenols at 320, 360 and 520 nm, respectively.

2.9. Analysis of Antimicrobial Activity in Dealcoholated Red Wines (DRW)

The biological materials used for the antimicrobial tests were strains of bacteria and yeasts: *Bacillus cereus* ŁOCK 0807, *Escherichia coli* ATTC 10536, *Listeria monocytogenes* ATTC 13932, *Salmonella Enteritidis* ATTC 13076, *Staphylococcus aureus* ATCC 6538 and *Candida albicans* ATCC 10231.

Antimicrobial Assay

The agar well diffusion method was used to verify whether the DRW affected the growth of the microorganisms [9]. A TSB (Trypticase soy broth) medium was used to activate cultures of bacteria following storage in the CRYOBANK™ system and YPD (Yeast Extract–Peptone–Dextrose) medium was used to activate the yeasts. After 24 h, the cultures were submitted for further analysis. Standardized inocula of the tested microorganisms were incubated in TSA (Trypticase soy agar) or YPD, depending on the groups of microorganisms. Next, wells with a diameter of 9 mm were punched aseptically with a sterile cork borer. To each well was added 120 µL of DRW. The samples were incubated at 37 °C for bacteria and 30 °C for yeast. After incubation, the inhibition zones were measured.

2.10. Statistical Analysis

Mean values, standard deviations and the occurrence of statistically significant differences were determined using STATISTICA 10 PL software (StatSoft, Krakow, Poland). The ANOVA test was used, assuming a significance level of 0.05.

3. Results and Discussion

3.1. Sugars, Organic Acids and Alcohols

The alcohol content of the wines ranged from 7.22% to 14.59% (*v/v*, Table 1). The lowest alcohol content was detected in the cranberry wines and the highest in the elderberry wines. The wines with the highest alcohol contents had low or trace amounts of glucose.

Table 1. Organic acids, sugars, glycerol (g/L) and ethanol (*v/v* in the investigated wines).

	BB	B	C	E	R	S
citric acid	1.17 ± 0.15 [a]	4.53 ± 0.35 [d]	1.95 ± 0.09 [b]	2.45 ± 0.08 [c]	8.74 ± 0.65 [f]	5.03 ± 0.15 [e]
malic acid	0.65 ± 0.05 [b]	2.03 ± 0.09 [f]	1.63 ± 0.07 [e]	1.15 ± 0.06 [d]	0.44 ± 0.02 [a]	0.95 ± 0.03 [c]
succinic acid	0.71 ± 0.05 [b]	0.99 ± 0.08 [c]	0.76 ± 0.05 [b]	0.92 ± 0.06 [c]	0.62 ± 0.03 [a]	0.56 ± 0.03 [a]
lactic acid	0.28 ± 0.02 [b]	0.02 ± 0.00 [a]	0.02 ± 0.00 [a]	0.54 ± 0.02 [c]	0.02 ± 0.00 [a]	0.02 ± 0.00 [a]
acetic acid	0.50 ± 0.03 [c]	0.24 ± 0.02 [b]	0.09 ± 0.01 [a]	uLOQ	uLOQ	uLOQ
ascorbic acid	0.08 ± 0.01 [c]	0.04 ± 0.00 [b]	0.03 ± 0.00 [a]	uLOQ	0.07 ± 0.00 [c]	uLOQ
glucose	uLOQ	uLOQ	43.43 ± 2.01 [c]	0.04 ± 0.00 [a]	0.10 ± 0.01 [b]	uLOQ
fructose	0.47 ± 0.03 [a]	0.68 ± 0.05 [b]	29.06 ± 1.98 [e]	0.67 ± 0.05 [b]	0.85 ± 0.06 [c]	1.62 ± 0.08 [d]
glycerol	8.58 ± 0.65 [c]	10.81 ± 0.77 [d]	4.95 ± 0.32 [a]	8.03 ± 0.63 [bc]	5.41 ± 0.35 [a]	6.96 ± 0.54 [b]
ethanol	12.86 ± 1.01 [c]	14.39 ± 1.06 [d]	7.22 ± 0.56 [a]	14.59 ± 0.99 [d]	8.43 ± 0.65 [b]	11.32 ± 0.87 [c]

BB—bilberry wine; B—blackberry wine; C—cranberry wine; E—elderberry wine; R—raspberry wine; S—strawberry wine. uLOQ—uder limit of quantification; Different letters in rows indicate a significant difference (*p* < 0.05).

Glycerol was the main by-product in wines obtained in our study. Its content was between 4.95 and 10.81 g/L. It was also the main compound in five of the eight raspberry wines investigated by Duarte et al. [37], with contents ranging from 4.6 to 10.2 g/L. In subsequent research by Duarte et al. [38], glycerol was again the main compound in 11 out of the 16 studied raspberry wines. The glycerol contents were similar to those reported in their previous study, in the range of 4.45–10.11 g/L.

Acidity is one of the most important parameters in wine. In grape wines, it is mainly associated with organic acids such as tartaric, malic, acetic and lactic acids. Citric acid can have a significant effect on fruit wines, due to its high concentration in the raw material (for example in raspberries and blackberries). Malic acid also contributes to the acidity of fruit wine. Lactic and succinic acids are produced during fermentation. In our study, these two compounds were produced in amounts lower than 1 g/L. The predominant volatile acid is acetic acid, often expressed as the wine-quality parameter and known as volatile acidity. It is always formed during alcoholic fermentation. Higher concentrations of acetic acid can affect the organoleptic properties of wine. The acetic acid content was under the limit of quantification in the elderberry, raspberry, and strawberry wines. The highest content was observed in the bilberry wines.

Duarte et al. found a high content of succinic acid in raspberry wines [37]. In only one sample was it below 5.0 (2.8 g/L), whereas, in the others, it was from 5.6 to 7.1 g/L [38]. This is almost ten times higher than the levels of succinic acid found in our raspberry wines. Duarte et al. [38] also reported higher acetic acid contents, from 0.7 to 2.3 g/L. Most of the wines analyzed had no glucose.

In our elderberry wines, citric acid was the main acid, at concentrations of around 2.40 g/L. Malic acid was present in the next highest quantities (1.15 g/L). This is consistent with results reported by Veberic et al. [39] for two cultivars and three selections of black elderberries from Slovenia. These authors found four acids (citric, malic, shikimic and fumaric) in the fruit.

Citric acid was present in the largest quantity, the content of malic acid was around three times smaller, and shikimic and fumaric acids composed on average 10% of the total.

Four of the six tested wines showed the presence of ascorbic acid. The concentration ranged from 0.03 to 0.08 g/L, for cranberry and bilberry wines, respectively. The low levels of vitamins in grape wines may explain the lack of relevant studies regarding the content of vitamins in fruit wines. Grape wines are reported to contain some B vitamins and very small amounts of vitamin C and fat-soluble vitamins [26]. Vitamin C can be destroyed during processing and storage.

3.2. Total Content of Polyphenols and Antioxidant Activity of Fruit Wines

Of the wines investigated in our study, wines made from elderberry fruits showed the highest content of total polyphenols (1480.47 mg/L) (Table 2). The content of total polyphenols in the remaining wines ranged from 408.03 to 759.42 mg GAE/L. Low concentrations of polyphenols were observed in wines from cranberries and bilberries (below 500 mg/L).

Table 2. Total polyphenols (mg/L) and antioxidant activity (mM of Trolox equivalents) of the investigated wines.

	BB	B	C	E	R	S
TP	466.82 ± 40.03 [a]	759.42 ± 52.30 [c]	408.03 ± 38.75 [a]	1480.47 ± 103.55 [d]	566.75 ± 43.02 [b]	525.60 ± 50.74 [b]
ABTS	4.29 ± 0.34 [c]	5.84 ± 0.34 [d]	3.03 ± 0.19 [a]	4.22 ± 0.28 [c]	3.49 ± 0.23 [b]	3.40 ± 0.18 [a]
DPPH	2.47 ± 0.19 [c]	2.55 ± 0.16 [d]	1.12 ± 0.07 [a]	1.87 ± 0.09 [b]	1.75 ± 0.07 [b]	1.66 ± 0.12 [b]
FRAP	5.07 ± 0.23 [c]	6.45 ± 0.45 [d]	3.32 ± 0.25 [a]	5.07 ± 0.39 [c]	4.42 ± 0.36 [b]	3.47 ± 0.22 [a]

BB—bilberry wine; B—blackberry wine; C—cranberry wine; E—elderberry wine; R—raspberry wine; S—strawberry wine; tr—traces; TP—total polyphenols; ABTS—2,2-azino-bis-3-ethylbenzothiazoline-6-sulfonic acid; DPPH—2,2-diphenyl-1-picrylhydrazyl; FRAP—Ferric reducing antioxidant power. Different letters in rows indicate a significant difference ($p < 0.05$).

In a study by Rupasinghe and Clegg [6], elderberry wines from Canada were found to contain 1753 mg GAE/L. Aged elderberry wines from Slovenia analyzed by Schmitzer et al. [7] contained 1584.99 mg/L of total polyphenols. The pitchings (musts) were prepared in a similar manner to our elderberry wines, and the results are similar.

Most studies that have measured the content of total polyphenols in blackberry and bilberry wines used commercial products, so the methods of preparation are not known. Blackberry and bilberry wines from Illinois have been reported as having total polyphenol contents ranging from 188 mg GAE/L to 1115 mg GAE/L [40]. These levels are comparable to those for the blackberry and bilberry wines in our study. The blackberry and bilberry wines investigated by Kalkan Yildrim [41] had slightly higher concentrations of polyphenols. The blackberry wines studied by Ortiz et al. [42] contained from 1122 to 1400 mg/L of total polyphenols (depending on the pectinolytic enzyme used). The blackberry wines analyzed by Ljevar et al. [43] contained from 1055 to 2704 mg/L. The highest contents of total polyphenols in blackberry wines have been reported by Mudnic et al. [44] (1697–2789 mg/L) and Mitic et al. [45] (1608–2836 mg/L).

The total polyphenol contents reported by Mitic et al. [45] and Ljevar et al. [43] for raspberry wines from Serbia and Croatia, respectively, were up to three-fold higher than those found in the raspberry wines in our study (1052–1490 and 1199–1840 mg/L, respectively). However, the preparation method again was different [46]. The raspberry and cranberry wines investigated by Rupasinghe and Clegg [6] contained 977 and 971 mg/L of total polyphenols, respectively.

The total content of polyphenols in our strawberry wines was almost three times lower than in wines obtained by Cakar et al. [1].

One method is usually insufficient to evaluate the antioxidant activity of a complex substance. In our study, three different assays were used to study the antioxidant properties of the wines. The antioxidant activity was determined in relation to ABTS$^+$, DPPH and FRAP. Significant differences between ABTS and DPPH radicals were found in the examined wines, but similar trends were observed in different assays (FRAP, ABTS, and DPPH). The highest values were observed in blackberry wines, and the lowest for the cranberry wines (Table 2). As in a study by Ljevar et al. [43], the blackberry wines in our research also showed the highest antioxidant activity. The elderberry wines, despite having the highest polyphenol content, did not show high antioxidant activity. Heinonen et al. [5] evaluated the antioxidant activity of over 44 different fruit wines, mainly berry wines. The results showed that the total phenolic content did not correlate with the antioxidant activity. On the other hand, some studies have confirmed a strong positive correlation between the total antioxidant activity in fruit wines and total phenolics [6,46].

A study by Gao et al. [47] investigating the contribution of three different antioxidant fractions using an ABTS assay showed the total antioxidant capacity of phenolics, ascorbic acid, and lipophilic compounds to be slightly lower than those of crude extracts. The phenolic fraction made a major contribution to the total activity (about 75%), followed by ascorbic acid (around 17%). According to Brand-Williams et al. [48], ascorbic acid is one of the fastest reacting antioxidants. Observing changes in the phenolic profile during the winemaking process, Lingua et al. [49] noted that anthocyanins were the most important phenols in the wine samples. Most berry wines are rich in anthocyanins, but they react poorly in the Folin–Ciocalteu test, giving a poor correlation [50]. In general, different compounds were selected to correlate with the different in vitro assays. Depending on the chemical structure of the compounds and the mechanisms involved (hydrogen atom transfer, single electron transfer, reducing power, and metal chelation, among others), they react differently in various vitro assays [51].

3.3. Polyphenols in Wines

As there is no extensive literature on the subject, qualitative and quantitative analysis of the composition of polyphenols in the fruit wines was based mainly on the data concerning the juices and fruits from which the wines originated.

3.3.1. Anthocyanins

Cyanidin 3-glucoside was present in all the samples (Table 3; Figures S1, S4, S7, S10, S13 and S16). The highest level of cyanidin-glucoside was found in the blackberry wines (30.26 mg/L). Cyanidin 3-galactoside was found in four of investigated wines. Traces of cyanidin 3-rutinoside were found in the blackberry and raspberry wines.

Table 3. Content of anthocyanins in the investigated wines (mg/L).

Compound	[M + H]$^+$ m/z	MS^2 m/z	BB	B	C	E	R	S
Cy-gal	449	287	4.11 ± 0.32	1.11 ± 0.08	0.65 ± 0.05	-	-	0.32 ± 0.03
Cy-glc	449	287	LOQ	30.26 ± 2.89	0.29 ± 0.02	6.19 ± 0.57	3.29 ± 0.28	0.30 ± 0.03
Cy-ara	419	287	-	-	1.19 ± 0.09	-	-	-
Cy-xyl	419	287	-	1.34 ± 0.09	-	-	-	-
Cy-rut	595	287	-	LOQ	-	-	LOQ	-
Cy-soph	611	287	-	-	-	-	14.98 ± 1.32	-
Cy-3(2glc) rut	757	287	-	-	-	-	1.53 ± 0.11	-
Cy-3mal-glc	535	287	-	1.32 ± 0.12	-	-	-	-
Cy-6mal-glc	535	287	-	13.19 ± 1.09	-	-	-	-
Cy-sam	579	537, 357	-	-	-	46.51 ± 4.34	-	-
Cy-sam-5-glc	744	287	-	-	-	23.07 ± 2.02	-	-
Cy-dioxalyl glc *	593/594	581, 287	-	6.50 ± 0.56	-	-	-	-
$\sum$ *Cy-deriv*			4.11	53.73	2.13	75.77	19.81	0.62
Dp-gal	465	303	5.02 ± 0.42	-	-	-	-	-
Dp-glc	465	303	0.81 ± 0.06	-	-	-	-	-
Dp-ara	435	303	0.63 ± 0.05	-	-	-	-	-
$\sum$ *Dp-der*			6.46	-	-	-	-	-
Mv-gal	493	331	1.01 ± 0.09	-	-	-	-	-
Mv-glc	493	331	8.81 ± 0.78	-	-	-	-	-
Mv-ara	463	331	1.02 ± 0.09	-	-	-	-	-
$\sum$ *Mv deriv*			10.84	-	-	-	-	-
Pg-glc	433	271	-	1.20 ± 0.09	-	-	-	1.29 ± 0.09
Pg-rut	579	433, 271	-	-	-	-	-	0.67 ± 0.06
Pg-3-acetyl-glc	475	271	-	-	-	-	-	0.49 ± 0.04
Pg-3mal-glc	519	271	-	-	-	-	-	0.35 ± 0.02
Pg-3,5diglc	595	433, 271	-	-	-	-	-	0.24 ± 0.02
Pg-3glc-rut	742/739		-	-	-	-	0.42 ± 0.03	-
E-(4,8)-Pg-glc	721	559	-	-	-	-	-	0.25 ± 0.02
(epi)afzelechin-Pg-glc	705	543, 407, 313	-	-	-	-	-	0.25 ± 0.02
CP Pg-glc	501	339	-	-	-	-	-	0.24 ± 0.02

Table 3. *Cont.*

Compound	[M + H]$^+$ m/z	MS^2 m/z	BB	B	C	E	R	S
$\sum Pg$ *deriv*			-	1.2	-	-	0.42	4.05
Pn-gal	463	301	0.65 ± 0.05	-	0.71 ± 0.06	-	-	-
Pn-glc	463	301	0.75 ± 0.06	0.82 ± 0.07	LOQ	-	-	-
Pn-ara	433	301	-	-	0.63 ± 0.05	-	-	-
$\sum$ *Pn deriv*			1.40	0.82	1.34	-	-	-
Pt-gal	479	317	0.49 ± 0.03	-	-	-	-	-
Pt-glc	479	317	8.72 ± 0.72	-	-	-	-	-
Pt-ara	449	317	1.29 ± 0.09	-	-	-	-	-
$\sum$ *Pt deriv*			10.50	-	-	-	-	-
ni	641	623, 505, 477, 605, 337	-	-	-	0.56 ± 0.04	-	-
total **			37.03	56.37	3.47	76.33	22.32	6.09

* or cyanidin 3-O-β-(6"-(3-hydroxy-3-methylglutaroyl)glucoside); ** of all peaks with max about 520 nm; Cy—cyanidin; Dp—delphinidin; Mv—malvidin; Pg—pelargonidin; Pn—peonidin; Pt—petunidin; gal—galactoside; glc—glucoside; ara—arabinoside, xyl—xyloside, rut—rutinoside, soph—sophoroside; sam—sambubioside; mal—malonyl, E—epicatechin; BB—bilberry wine; B—blackberry wine; C—cranberry wine; E—elderberry wine; R—raspberry wine; S—strawberry wine; LOQ—under limit of quantitation.

Cyanidin-sophoroside and cyanidin 3-(2G-glucosylrutinoside) were found in the raspberry wines, with Cy-soph as the main compound (Figure S13). Cyanidin 3-sambubioside and cyanidin 3-sambubioside-5-glucoside were present only in the elderberry wines, in quantities of 46.51 mg/L and 23.07 mg/L, respectively (Figure S10).

Delphinidin derivatives were only present in the bilberry wines (Figure S1). Malvidin derivatives were also detected only in the bilberry wines, whereas pelargonidin derivatives were found in the wines made from blackberries and strawberries (Figures S4 and S16). Pelargonidin 3-glucoside was the main anthocyanin found in the strawberry wines. Carboxypyranopelargonidin-glucoside (CP Pg-glc) was also found in the strawberry wines. This compound had been identified previously in strawberries [52] and in strawberry-fermented products [11,53].

Peonidins were found in the bilberry, blackberry and cranberry wines. Petunidins were found only in the bilberry wines.

The highest concentrations of anthocyanins were found in the elderberry wines (76.33 mg/L). The concentrations of anthocyanins in the blackberry wines were in excess of 50 mg/L.

The main compound present in elderberry wines was cyanidin 3-sambubioside (46.51 mg/L), and cyanidin 3-sambubioside-5-glucoside was present in the next largest quantities.

Our results differ significantly from those reported by Schmitzer et al. [7]. In their study of elderberry wines, cyanidin 3-glucoside was present at the highest concentrations in mature wine (20.88 mg/L), whereas, in our study, the concentration of this compound was 6.19 mg/L. According to Schmitzer et al., cyanidin 3,5-diglucoside was present at a slightly lower concentration (18.49 mg/L), whereas, in our study, this compound was not found in the elderberry wines. Cyanidin 3,5-diglucoside and cyanidin 3-sambubioside-5-glucoside have similar retention times and coelution sometimes occurs [54]. However, no compound with a mass characteristic of cyanidin 3,5-diglucoside was found during our investigation. The sum of the concentrations of anthocyanins observed by Schmitzer et al. [7] based on HPLC analysis was approximately 50% higher than that in our elderberry wines.

The dominant compound in the blackberry wines we studied was cyanidin 3-glucoside, at a concentration of 30.26 mg/L. In domestic wines obtained by Mitic et al. [45], the content of cyanidin 3-glucoside was around 10 times higher. Cyanidin xyloside was present in the next highest concentration, which was again much higher than in the blackberry wines in our study. Cyanidin 3-rutinoside was the main compound in the blackberry wines analyzed by Ljevar et al. [43]. Only small amounts of this compound were found in the blackberry wines we studied. Relatively large amounts of cyanidin 3-glucoside acylated with malonic acid were detected in our study, at a concentration approximately two-fold lower than those for cyanidin 3-glucoside. This compound has been identified previously as occurring in blackberries [54–58] but was identified here for the first time in blackberry wine.

Compared to those analyzed by Hornedo-Ortega et al. [53], the strawberry wines in our study contained much lower levels of both pelargonidin 3-glucoside and pelargonidin 3-rutinoside. However, the wines studied by Hornedo-Ortega et al. were analyzed immediately following the fermentation process, whereas ours were tested after 5 months of aging. Other authors have noted a 63–85% reduction in these compounds during the fermentation and storage [59,60]. Relatively large amounts of one acylated derivative and one diglucoside derivative of pelargonidin were found in our wines. Hornedo-Ortega et al. [53] showed that fermentation significantly increases the levels of diglucoside in strawberry wines (2.5- and 6.2-fold for 2012 and 2013 vintages, respectively).

All of the wines in our study were prepared using thermal treatment. As observed by Behrends and Weber [10], pre-fermentative heat treatment influences the characteristics of wine. This was confirmed in our previous research [2,3]. The bilberry wines investigated by Behrends and Weber [10] had almost the same anthocyanin profiles as juices when pre-fermentative heat treatment was applied. With warm treatment (70 °C), the ratio of glucosides to galactosides and arabinosides was 47.9:32.1:20. In the bilberry wines we studied, the ratio was 58.9:32:9.1. Some of the fruits in our study (including bilberries) were also treated with pectinase. As observed by Buchert et al. [61], enzyme-assisted bilberry juice production leads to greater losses of galactosides.

3.3.2. Phenolic Acids

The largest quantities of hydroxycinnamic acids were found in elderberry wines (150.79 mg/L). The lowest levels were observed in the strawberry wines (13.43 mg/L) (Table 4). None of the acids were present in all of the wines (Figures S2, S5, S8, S11, S14 and S17).

The main acids found in the elderberry wines were neochlorogenic, chlorogenic and caffeic acids (Figure S11). Schmitzer et al. [7] analyzed only neochlorogenic and chlorogenic acids. The neochlorogenic acid content was higher in our samples. It is difficult to compare results for chlorogenic acid. During chromatographic analysis (HPLC), chlorogenic acid probably coeluted with caffeic acid and one large peak was observed. Using a mass spectrometer, a mass of 179 (typical for caffeic acid) proved dominant. Caffeic acid hexoside (341 *m/z*), *p*-coumaric acid (163 *m/z*) and p-coumaric acid derivative (525 *m/z*) were also found. Caffeic acid and its derivative, caffeic acid hexoside, were the main acids in the bilberry wines, accounting for almost 75% of the total acid content (Figure S2). Caffeic acid was also the main compound in the raspberry wines. The dominant acids in the blackberry wines (around 85%) and strawberry wines (around 47%) were p-coumaroylhexosides. Cakar et al. [1] found small amounts (4.16–2.83 µg/mL) of p-coumaric acid in strawberry wines, which was about 10% of the total phenolic acid content. These authors identified chlorogenic acid chlorogenic acid as a main compound (290–335 µg/mL) followed by the caffeic acid. No chlorogenic acid was identified in any of the 90 strawberry varieties investigated by Nowicka et al. [62]. A compound with very similar MS (*m/z* 355) was identified as 1-*O*-trans-cinnamoyl-glucose. This compound had the highest content in almost all investigated varieties, followed by *p*-coumaroyl-glucosides. Nowicka et al. [62] also identified 1-*O*-feruloylglucose. In our investigated wines, 5-hydroxyferuloylhexoside was identified. This compound probably formed during fermentation.

3.3.3. Flavonols

Flavonols were present in relatively low concentrations compared to the other groups of studied compounds, from 0.87 to 9.27 mg/L (Table 5). The lowest concentrations occurred in wines made from raspberries, strawberries, bilberries and blackberries (0.87–1.81 mg/L), and the highest in wines made from elderberries (9.27 mg/L).

Table 4. Content of hydroxycinnamic acid derivatives in the investigated wines (mg/L).

	$[M - H]^-$ m/z	MS^2 m/z	BB	B	C	E	R	S
malonylo-CQA	439, 396	395, 219, 173, 295, 289	-	-	-	-	-	0.74 ± 0.05
neoChA	353	191, 179	-	6.42 ± 0.56	-	24.99 ± 2.08	-	-
CAH	341	197, 135, 161, 179	10.97 ± 0.93	LOQ	10.89 ± 0.87	9.98 ± 0.78	4.41 ± 0.37	-
pCoH	325	163, 145, 187, 265	-	54.35 * ± 4.89	-	-	12.35 * ± 0.96	6.31 ± 0.54
ChA	353	191, 179	LOQ	2.01 ± 0.17	5.59 ± 0.45	coeluted	-	-
CA	179	135	44.10 ± 3.99		8.67 ± 0.78	60.04 ± 5.33	15.32 ± 1.03	
FA	193	134	-	-	-	-	-	LOQ
p-CoA	163	119	14.77 ± 1.23	-	-	13.44 ± 1.02	-	4.60 ± 0.34
pCo der	411		2.27 ± 0.19	-	-	-	-	-
pCo der	525		-	-	-	19.86 ± 1.88	-	-
5-hydroxy F hex	371	281, 251, 221, 209	-	-	-	-	-	1.78 ± 0.16
ni	207		-	-	-	16.20 ± 1.52	-	-
total			73.48	63.95	37.06	150.79	32.18	13.43

- not identified in this wine; malonylo-CQA-malonylo-caffeoylquinic acid; neoChA-neochlorogenic acid; CAH—caffeic acid hexoside; ChA—chlorogenic acid; pCoH—p-coumaroylhexoside; CA—caffeic acid; FA—ferulic acid; pCoA—p-coumaric acid; pCoA der—p-coumaric acid derivative; 5-hydroxyFhex—5-hydroxyferuloyl hexose; tr—traces; BB—bilberry wine; B—blackberry wine; C—cranberry wine; E—elderberry wine; R—raspberry wine; S—strawberry wine; LOQ—under limit of quantitation; *—two peaks.

Table 5. Flavonols contents in the investigated wines (mg/L).

	[M − H]⁻ m/z	MS MS² m/z	BB	B	C	E	R	S
M-glc	479	317	0.27 ± 0.02	-	-	-	-	-
M-ara	449	317	-	-	0.01 ± 0.00	-	-	-
M-xyl	449	317	-	-	0.02 ± 0.00	-	-	-
M-malonylglc	565	317	-	-	-	-	-	LOQ
M-dimethoxy-hex	507	344, 387	-	-	0.01 ± 0.00	-	-	-
M	317	179, 151, 192	0.28 ± 0.02	-	0.78 ± 0.06	-	-	-
∑ M derivatives			0.55	-	0.82	-	-	LOQ
Q-gal	463	301	-	0.36 ± 0.03	0.05 ± 0.00	0.10 ± 0.01	0.11 ± 0.01	-
Q-glc	463	301	0.17 ± 0.01	0.07 ± 0.01	-	2.36 ± 0.18	0.05 ± 0.00	0.19 ± 0.02
Q-ara	433	301	-	-	0.12 ± 0.01	-	-	-
Q-rut	609	301, 343, 463	LOQ	0.09 ± 0.00	-	5.04 ± 0.42	0.10 ± 0.01	-
Q-pent	433	301, 179, 151	-	LOQ	-	-	-	-
Q-xyl	433	301	0.05 ± 0.00	-	0.20 ± 0.02	-	-	-
Q-rha	447	301	-	-	0.25 ± 0.02	-	-	-
Q-gluc	477	301	-	0.31 ± 0.02	-	-	0.09 ± 0.01	0.38 ± 0.03
Q-diglc	625	283, 255, 463, 301	LOQ	-	-	-	-	-
Q-2gal-rha	609	283, 255, 300	-	-	-	-	0.09 ± 0.01	-
Q-3acetylhex	505	463, 301	-	0.39 ± 0.03	-	-	-	LOQ
Q-methoxyhex	493	463, 301	0.11 ± 0.01	-	-	-	-	-
Q3[6″(3hydroxy-3 methyl-glut)] gal	607	463, 301	-	0.51 ± 0.04	-	-	-	-
Q-malonyl-glc	549	503, 301	-	-	-	LOQ	-	-
methoxyQ-xyl	447	300	-	-	0.01 ± 0.00	-	-	-
Q-benzoyl gal	567	300	-	-	0.05 ± 0.00	-	-	-
Q	301	179, 151, 257	0.28 ± 0.02	0.07 ± 0.00	1.56 ± 0.14	1.35 ± 0.12	0.04 ± 0.00	0.05 ± 0.00
IsoQ	509	463	-	-	-	LOQ	-	-
dihydroQ glc	465	285, 151	-	LOQ	-	-	-	-
I=3-methylQ	315	631/632, 315	-	-	-	0.19 ± 0.02	-	-
∑ Q derivatives			0.63	1.80	2.24	9.04	0.48	0.62

Table 5. *Cont.*

	$[M-H]^-$ *m/z*	MS MS^2 *m/z*	BB	B	C	E	R	S
K-gal	447	285	0.16 ± 0.01	-	-	-	0.10 ± 0.01	-
K-glc	447	285	0.16 ± 0.01	-	-	-	-	0.15 ± 0.01
K-rut	593	285	-	-	-	0.22 ± 0.02	-	-
K-pent	417	241, 152, 285	-	-	-	-	-	LOQ
K-gluc	461	415, 285	-	-	-	-	0.26 ± 0.02	0.15 ± 0.02
K	285	267	-	-	-	-	0.03 ± 0.00	-
dihydroK-glc	449	431, 287, 269, 259, 243, 179	-	LOQ	-	-	-	LOQ
dihydroK-rha	433	287	-	LOQ	-	-	-	-
K3(6″-*p*-Co)glc	593		-	LOQ	-	-	-	-
∑ K derivatives			0.16	-	-	0.22	0.39	0.30
total			1.43	1.81	3.02	9.27	0.87	0.92

- not identified in this wine; M—Myricetin; Q—Quercetin; K—Kaempferol; I—Isorhamnetin; gal—galactoside; glc—glucoside; ara—arabinoside, xyl—xyloside, rut—rutinoside, soph—sophoroside; rha—rhamnoside; glu—glucuronide; glut—glutaroyl; p-Co—p-coumaroyl; pent—pentoside; hex—hexoside; BB—bilberry wine; B—blackberry wine; C—cranberry wine; E—elderberry wine; R—raspberry wine; S—strawberry wine; LOQ—under limit of quantitation.

In terms of qualitative composition, more than 80% of all identified flavonols in the elderberry wines were quercetin and its derivatives (Figure S12). Myricetin and its derivatives were found in the cranberry and bilberry wines (Figures S3 and S9), while trace amounts of myricetin malonylglucoside were found in strawberry wines. Quercetin 3-rutinoside was the main flavonol in the elderberry wines, comprising 56% of the total for this group of compounds. Flavonols were found to dominate in the elderberry wines studied by Schmitzer et al. [7], at concentrations around 10 times higher than in our wines, with the highest concentrations occurring in young wines. The same authors detected quercetin rutinoside and glucoside in the highest and second-highest concentrations. Kaempferol-rutinoside was present in the lowest quantity, at a concentration of 1.1 mg/L, whereas, in our elderberry wines, the concentration was 0.22 mg/L. The elderberry wines in our study contained more than 1 mg/L of quercetin. Fruit wines made from cherries, blackberries and raspberries were found by Mitic et al. [45] to contain derivatives of quercetin and kaempferol. The same authors reported the presence of kaempferol derivatives in blackberry wines, at levels of 0.30–0.85 mg/L. However, we did not detect these compounds in our blackberry wines (Figure S6). The content of quercetin derivatives in the raspberry wines studied by Mitic et al. [45] ranged from 0.98 to 1.80 mg/L, compared with only 0.48 mg/L in our study. We also found kaempferol derivatives in raspberry wines (Figure S15), at a concentration of 0.39 mg/L. The strawberry wines investigated by Cakar et al. [1] contained three compounds from the group of flavonols: quercetin, quercetin 3-rutinoside (rutin) and kaempferol. Only small amounts of quercetin were found in our strawberry wines, less than 10% of the total Q-derivatives content. We did not identify Q-rut, but we found significant amounts of Q-glucoside and Q-glucuronide (Figure S18). We were unable to identify kaempferol in the strawberry wines, but we did find its derivatives, K-glucoside and K-glucuronide, as well as traces of K-pentoside and dihydroK-glucoside.

3.3.4. Other Bioactive Compounds

Other bioactive compounds were tentatively identified using LC–MS^n (Table 6). Sixteen acids were found, including hydroxybenzoic acids. Cinnamic acid was identified in all the samples. Shikimic acid was found in all of the wines except cranberry wine. In the blackberry wines, two forms of abscisic acid were identified: abscisic acid D-glucopyranosyl ester (ABA-GE) and ursolic acid.

Ellagic acid was identified in five of the six wines in our study, but its derivatives were found mainly in the wines made from blackberries (five compounds), strawberries (three compounds) and raspberries (two compounds). Of the twelve ellagitannins we identified, as many as ten were found in the blackberry wines. Five were found in the strawberry wines and four were identified in the raspberry wines. Three gallic acid derivatives were identified: one in the bilberry wines, one in the cranberry wines and one in the strawberry wines. Procyanidins are another important group of polyphenols. In total, 20 compounds from this group were identified. Of these, 14 were found in the cranberry wines, which were the richest and most diverse source of procyanidins.

Seven compounds from this group were identified in the strawberry wines, including two afzelechin-catechin derivatives.

Cakar et al. [1] identified hydroxybenzoic acids in strawberry wines. Gallic acid had the highest concentration, followed consecutively by p-hydroxybenzoic, protocatechuic and vanillic acids. We identified two of these hydroxybenzoic acids in our strawberry wines (Table 6). We also found two derivatives of: protocatechuic acid hexoside and 1-*O*-protocatechuylhexoside. Vanillic acid was not found.

Cakar et al. also found significant amounts of ellagic acid in their strawberry wines. We identified this compound in five wines (was not present in cranberry wines) (Table 6).

Table 6. Other compounds identified in investigated wines.

Tentative Compound	λmax	[M − H]⁻ m/z	MS² m/z	BB	B	C	E	R	S
acids									
cinnamic acid	225	147	129, 85, 87, 103	+	+	+	+	+	+
vanillic acid	271	167		-	-	+	-	-	-
ascorbic acid	253	175	129, 115, 157, 85	+	+	+	-	+	-
shikimic acid	270	173	127, 83	+	+	-	+	+	+
p-hydroxybenzoic acid	277	137	93, 119, 110	-	-	+	+	+	+
benzoic acid	275	121	77, 121, 92	-	-	+	-	-	-
hydroxybenzoyl-glc	276, 309	299	137	-	-	-	-	-	+
protocatechuic acid	260, 294	153	109, 125, 83	+	+	+	+	-	-
protocatechuic acid hex		315	152, 108	-	+	+	-	-	+
1-O-protocatechuylhex		285	152, 108	-	-	+	-	-	+
sinapic acid hex	265, 382	385	339	-	-	-	+	-	-
brevifolin carboxylic acid	281	291	248, 247, 203	-	-	-	-	-	+
cis-ABA		263	153	+	+	-	-	-	-
trans-ABA		263	204	+	+	-	-	-	-
ABA-GE		425	263	-	+	-	-	-	-
ursolic acid=prunol		455	515	-	+	-	-	-	-
Ellagic acid derivatives									
ellagic acid	245, 278, 382	603 [2M] 301	467, 439, 179, 273, 257	+	+	-	+	+	+
ellagic acid pent	231	433	300/301	-	+	-	-	+	+
ellagic acid hex	255, 362	463	301	-	+	-	-	-	-
ellagic acid deoxyhex	231, 364	447	300/301, 257	-	-	-	-	-	+
dimethyl ellagic acid pent		461	300/301, 145	-	+	-	-	-	-
ellagic acid acetyl-ara	235, 273	475	301	-	-	-	-	+	-
methylellagic acid gluc	253, 361	491	315, 301, 257, 229	-	+	-	-	-	-
ellagic acid acetyl-methylpent	254, 364	489	301, 257, 229	-	+	-	-	-	-
ellagic acid rha		447	301	-	-	-	-	-	+

Table 6. *Cont.*

Tentative Compound	λmax	[M − H]⁻ m/z	MS² m/z	BB	B	C	E	R	S
Ellagitannins									
ellagitannin	232, 270	679	664	-	-	-	-	-	+
HHDP glc		481	301, 275	-	+	-	-	-	+
galloyl-bis-HHDP glc		935	633, 301	-	+	-	-	-	-
galloyl-HHDP glc	280	633, 632.6	481, 301, 613, 301, 481, 783	-	+	-	-	-	+
bis-HDDP-glc	280	783	301, 481, 257, 229	-	+	-	-	-	+
tris-galloyl-HHDP hex		951	907, 783, 605, 301	-	+	-	-	-	+
davuriicin M1 (diHHDP-glc-galloyl-ellagic acid)		617[M-2H]²⁻, 1236	933, 631, 301	-	+	-	-	-	-
Sanguiin H-10 isomer (2)	232	[1567]⁻, [783]²⁻	935, 633, 301	-	+	-	-	+	-
Sanguiin H-2	245	1103, [551]²	935, 633, 469, 301	-	-	-	-	+	-
castalagin/vescalagin		933	301	-	+	-	-	-	-
pedunculagin/sanguin isomer H10	268, 377	783	633, 301, 1266, 934, 1104	-	+	-	-	+	-
Sanguin H6	340, 352, 366	935/934 [M-2H]²⁻ 1870	633, 301, 897, 916, 783, 1567, 1235, 633, 301	-	+	-	-	+	-
Gallic acid derivatives									
gallic acid	286	169	125	+	-	-	-	-	+
methyl gallate		183/184		-	-	+	-	-	-
galloylquinic acid		343	191, 169	-	-	-	-	-	+
gallic acid deriv	280, 451	635	483	+	-	-	-	-	-
Procyanidins									
epigallocatechin	283	611		-	+	-	-	-	+
gallocatechin		306/305		-	+	-	-	-	-
catechin	280	289	245, 205, 179	+	+	+	-	-	+
epicatechin	285	289	245, 205, 271, 179	-	+	-	-	+	-
Procyanidin dimer	277	575	490, 499, 413	-	-	+	-	+	-

Table 6. *Cont.*

Tentative Compound	λmax	[M − H]⁻ m/z	MS² m/z	BB	B	C	E	R	S
ni	277	575	377, 395, 333, 273, 1007	-	-	+	-	-	-
	277	575	863/864, 499, 413, 267, 289, 699, 1025	-	-	+	-	-	-
	277	575	499, 490, 861, 423, 289, 999, 1025	-	-	+	-	-	-
	277	575	395, 351, 371, 289, 1025	-	-	+	-	-	-
	277	575	423, 449, 539, 285, 557, 1025	-	-	+	-	-	-
B type dimer (procyanidin dimer)	282	577	425, 407, 451	+	-	-	-	-	+
Procyanidin B1	278	577	425, 407	-	+	+	-	-	+
		577	397, 373, 273, 415, 1019	-	-	+	-	-	-
Procyanidin trimer (Atype)	280	863	711, 411, 559, 693	-	-	+	-	-	-
Procyanidin trimer (Btype)	281	865	695, 577, 407, 847	-	-	+	-	-	+
Procyanidin tetramer (Btype)	276	1152/1153		-	-	+	-	-	-
dimer (Cat-Afz) propelargonidin dimer	279	561	289, 543, 435	-	+	-	-	-	+
trimer A type	276	863	711, 693, 411, 459, 559, 289	-	-	+	-	-	-
ni		863	575, 711, 693, 559, 285, 1601	-	-	+	-	-	-
Trimer (Cat-Cat-Afz)		849		-	-	-	-	-	+
Flavone.									
apigenin pent		401	269, 161	-	+	-	-	-	-
apigenin glc		431	370, 269, 311	-	+	-	-	-	-
Biflavonoids									
pentahydroxyflavan dimer	250	579	271, 289	-	-	-	-	+	-
tetrahydroxyflavan–pentahydroxyflavan dimer		563	273, 291, 411, 427	-	-	-	-	+	-
Stilbenoids									
trans-resveratrol-glc		389	185, 227	+	+	-	-	-	-
Unknown compounds									
ni		340	294, 188, 161	-	-	-	+	-	-
ni	226, 278, 397	405	225	-	-	-	+	-	-
ni	259	391	217, 373, 111, 216, 191	-	-	-	-	+	-
ni	226, 284	379	241	-	-	-	+	-	-
ni	281	333	165, 289, 183	-	-	+	-	-	-

ni—not identified compound; +—present; - —absent; glc—glucoside, hex—hexoside, ABA—abscisic acid; ABA-GE—abscisic acid D-glucopyranosyl ester; BB—bilberry wine; B—blackberry wine; C—cranberry wine; E—elderberry wine; R—raspberry wine; S—strawberry wine.

3.4. Effect of Dealcoholated Fruit Wines on Microbial Growth

We studied the effects of the compounds present in the dealcoholated fruit wines on the growth of various microorganisms (Table 7). The berry wines had no inhibitory effect on the growth of *Salmonella* Enteritidis, *Staphylococcus aureus* bacteria and *Candida albicans* yeast. The only growth inhibitors for *Escherichia coli* ATCC 1053 were bioactive compounds found in the strawberry wines. The resulting zones of inhibition were 2.67 mm. The strain *Bacillus cereus* ŁOCK O807 was the most susceptible to the effects of the wines. Its growth was inhibited by the compounds in five of the wines. Only the elderberry wines had no effect on the growth of this strain. The most extensive inhibition zones resulted from the impact of raspberry wines.

Table 7. Inhibition zones (mm).

	Escherichia coli	*Salmonella* Enteritidis	*Bacillus cereus*	*Listeria monocytogenes*	*Staphylococcus aureus*	*Candida albicans*
Bilberry	-	-	1.73 ± 0.12 [A]	-	-	-
Blackberry	-	-	2.00 ± 0.25 [bA]	1.00 ± 0.00 [aA]	-	-
Cranberry	-	-	1.83 ± 0.23 [aA]	2.33 ± 0.58 [aB]	-	-
Elderberry	-	-	-	-	-	-
Raspberry	-	-	4.00 ± 0.71 [B]	-	-	-
Strawberry	2.67 ± 0.58 [a]	-	1.83 ± 0.23 [aA]	-	-	-

- No inhibitory effect; a,b—Different letters indicate a significant difference in rows ($p < 0.05$); A,B—Different letters indicate a significant difference in columns ($p < 0.05$).

Growth of *Listeria monocytogenes* ATCC 13932 was inhibited by the bioactive compounds present in two of the wines. The cranberry and blackberry wines had an inhibitory effect on the growth of these bacteria, with a zones of 2.33 and 1.00 mm, respectively.

The elderberry wines did not inhibit the growth of any of the microorganism, despite having the highest total polyphenol concentration and the highest content of anthocyanins, which some authors consider to be one of the main providers of antimicrobial properties [17,18]. The growth of *Bacillus cereus* was inhibited to the greatest extent by the wine made from raspberries. These fruits are a rich source of ellagitannins, which have strong antimicrobial activity. In the raspberry wines, we identified Sanguiin H2, H6 and Sanguiin H10 isomers (Table 6). Other compounds from this group (HHDP glucosides and their derivatives) were identified in the blackberry and strawberry wines. The cranberry wines did not contain ellagitannins, but a wide range of procyanidins were identified, including type A. Proanthocyanidin extracts of cranberries investigated by Kylli et al. [63] showed strong antimicrobial effects against *Staphylococcus aureus*, whereas they had no effect on other bacterial strains such as *Salmonella* Typhimurium and *Escherichia coli*. Phenolic extracts of lingonberry and cranberry had an antibacterial effect on Gram-positive pathogens including *Staphylococcus*, *Bacillus* and *Clostridium*, but only had weak or no antimicrobial activity on Gram-negative strains of *Salmonella*. However, *Listeria monocytogenes* was not inhibited by by either the lingonberry extract or cranberry extract [21,64].

Based on the results of these preliminary studies on the dealcoholated red berry wines, we see the possibility of using them as a food additive, improving safety and extending shelf life. As previously mentioned, some of the microorganisms tested may already be found in the oral cavity, so the wines could have an impact at this stage. Regarding the upper respiratory tract, oral invasion in immunosuppressed patients may be more frequent than previously documented, as the oral cavity can be colonized by *B. cereus* either by inhaling spores or by eating food contaminated with *B. cereus* [32]. Foci can occur when bacteria become trapped in the furrows in the oral cavity, where they grow and release toxins that spread to adjacent tissues and other parts of the body. However, further studies are necessary to investigate the action of the wines against pathogens in the human gastrointestinal tract.

4. Conclusions

In this study, about 150 compounds were identified in berry wines, including anthocyanins (34), hydroxycinnamic acids (12) and flavonols (36). Some of these compounds were identified for the first time in berry wines. The largest number of bioactive compounds was identified in the blackberry wines (59 compounds). All of the wines were rich in polyphenols. Elderberry wines were the richest source of polyphenols (over 1000 mg/L) and contained the largest amounts of all of analyzed groups of compounds (hydroxycinnamic acids, anthocyanins and flavonols). The lowest concentrations of polyphenols were found in wines made from cranberries and bilberries (below 500 mg/L). The dealcoholated berry wines were found to inhibit *Bacillus cereus* growth. Elderberry wines, despite their high content of polyphenols, did not show antimicrobial properties against the tested microorganisms. Antimicrobial properties may be affected by the combination and proportions of active compounds, and not only by the individual compounds. Our results show that berry fruit wines could provide biologically active compounds and at the same time protect against pathogens.

Supplementary Materials: The following are available online at http://www.mdpi.com/2304-8158/9/12/1783/s1, Figure S1: Chromatogram at 520 nm of bilberry wine obtained with HPLC-DAD; 3-Dp-gal; 4-Dp-glc; 5-Cy-gal; 6-Dp-ara; 7-Cy-glc; 8-Pt-gal; 9-Pt-glc; 10-Pn-gal;12-Pt-ara;13-Pn-glc; 14-Mv-gal; 15-Mv-glc; 16-Mv-ara. Figure S2: Chromatogram at 320 nm of bilberry wine obtained with HPLC-DAD; 7-CAH; 8-CA; 10-p-CoA; 18-p-CoA der. Figure S3: Chromatogram at 360 nm of bilberry wine obtained with HPLC-DAD; 11-M-glc; 14-Q-glc; 15-Q-metoxyhex; 18-K-gal; 19-K-glc; 21-M; 22-Q. Figure S4: Chromatogram at 520 nm of blackberry wine obtained with HPLC-DAD; 1-Cy-gal; 2-Cy-glc; 3-Cy-xyl; 5-Pg-glc; 7-Cy-3mal-glc; 8-Cy-6mal-glc; 9-Cy-dioxalyl glc. Figure S5: Chromatogram at 320 nm of blackberry wine obtained with HPLC-DAD; 11-CAH; 15-neoChA; 18-ChA; 21-pCoH. Figure S6: Chromatogram at 360 nm of blackberry wine obtained with HPLC-DAD; 23-Q-rut; 24-Q-gal; 25-Q-gluc; 26-Q-glc; 28-Qacetylhex; 29-Q-3[6″ (3hydroxy-3 methyl-glut)] gal. Figure S7: Chromatogram at 520 nm of cranberry wine obtained with HPLC-DAD; 10-Cy-gal; 11-Cy-glc; 14-Cy-ara; 15-Pn-gal; 17-Pn-glc; 18-Pn-ara. Figure S8: Chromatogram at 320 nm of cranberry wine obtained with HPLC-DAD; 11-CAH; 12-ChA; 13-CA. Figure S9: Chromatogram at 360 nm of cranberry wine obtained with HPLC-DAD; 7-M-xyl; 9-M-ara; 14-Q-gal; 17-M-dimetoxy-hex; 18-Q-xyl; 19-Q-ara; 20-Q-rha; 21-M; 22-metoxyQ-xyl; 23-Q; 24-Q-benzoyl gal. Figure S10: Chromatogram at 520 nm of elderberry wine obtained with HPLC-DAD; 3-Cy-sam-5-glc; 4-Cy-sam; 5-Cy-glc. Figure S11: Chromatogram at 320 nm of elderberry wine obtained with HPLC-DAD; 7-neoChA; 13-CAH; 16-CA/ChA -coeluted; 23-p-CoA der; 32-ni (λmax = 323). Figure S12: Chromatogram at 360 nm of elderberry wine obtained with HPLC-DAD; 17-Q-rut; 18-Q-glc;19-K-rut; 20-3-methylQ; 21-Q. Figure S13: Chromatogram at 520 nm of raspberry wine obtained with HPLC-DAD; 5-Cy-soph; 6-Cy-3(2glc) rut; 7-Pg-3glc-rut/Cy-glc. Figure S14: Chromatogram at 320 nm of raspberry wine obtained with HPLC-DAD; 12-CAH; 16, 17-pCoAHs; 18-CA. Figure S15 Chromatogram at 360 nm of raspberry wine obtained with HPLC-DAD; 14-Q-2gal-rha; 20-Q-rut; 21- Q-gluc; 22-K-gal; 23-K-gluc; 25-Q; 26-K. Figure S16: Chromatogram at 520 nm of strawberry wine obtained with HPLC-DAD; 7-Cy-gal; 12 Cy-glc; 15- Pg-glc; 16-Pg-rut; 17-Pg-3,5diglc; 18-Pg-3mal-glc; 22-Pg-3-acet-glc. Figure S17: Chromatogram at 320 nm of strawberry wine obtained with HPLC-DAD; 12-malonyloCQA; 19-p-CoH; 31-pCoA; 42-5-hydroxyF hex. Figure S18 Chromatogram at 360 nm of strawberry wine obtained with HPLC-DAD; 39-Q-gluc; 40-Q-glc; 41-K-gluc; 42-Q.

Author Contributions: Conceptualization, A.C. and A.W.; methodology, A.C., A.W., A.N.; investigation, A.C., A.W., A.S.; writing—original draft preparation, A.C.; writing—review and editing, A.C., A.N. All authors have read and agreed to the published version of the manuscript.

Funding: This research received no external funding.

Conflicts of Interest: The authors declare no conflict of interest.

References

1. Čakar, U.; Petrović, A.; Pejin, B.; Čakar, M.; Živković, M.; Vajs, V.; Đorđević, B. Fruit as a substrate for a wine: A case study of selected berry and drupe fruit wines. *Sci. Hortic.* **2019**, *244*, 42–49. [CrossRef]
2. Czyżowska, A.; Pogorzelski, E. Changes to polyphenols in the process of production of musts and wines from blackcurrants and cherries. Part I. Total polyphenols and phenolic acids. *Eur. Food Res. Technol.* **2002**, *214*, 148–154. [CrossRef]
3. Czyżowska, A.; Pogorzelski, E. Changes to polyphenols in the process of production of musts and wines from blackcurrants and cherries. Part II. Anthocyanins and flavanols. *Eur. Food Res. Technol.* **2004**, *218*, 355–359. [CrossRef]

4. Czyżowska, A.; Klewicka, E.; Pogorzelski, E.; Nowak, A. Polyphenols, vitamin C and antioxidant activity in wines from *Rosa canina* L. and *Rosa rugosa* Thunb. *J. Food Compos. Anal.* **2015**, *39*, 62–68. [CrossRef]

5. Heinonen, M.I.; Lehtonen, P.J.; Hopia, A.I. Antioxidant activity of berry and fruit wines and liquors. *J. Agric. Food Chem.* **1998**, *46*, 25–31. [CrossRef]

6. Rupasinghe, H.P.V.; Clegg, S. Total antioxidant capacity, total phenolic content, mineral elements, and histamine concentrations in wines of different fruit sources. *J. Food Compos. Anal.* **2007**, *20*, 133–137. [CrossRef]

7. Schmitzer, V.; Veberic, R.; Slatnar, A.; Stampar, F. Elderberry (*Sambucus nigra*) wine. A product rich in health promoting compounds. *J. Agric. Food Chem.* **2010**, *58*, 10143–10146. [CrossRef]

8. Vuorinen, H.; Maatta, K.; Torronen, R. Content of flavonols myricetin, quercetin, and kaempferol in Finnish berry wines. *J. Agric. Food Chem.* **2000**, *48*, 2675–2680. [CrossRef]

9. Wilkowska, A.; Czyżowska, A.; Ambroziak, A.; Adamiec, J. Structural, physicochemical and biological properties of spray-dried wine powders. *Food Chem.* **2017**, *228*, 77–84. [CrossRef]

10. Behrends, A.; Weber, F. Influence of different fermentation strategies on the phenolic profile of bilberry wine (*Vaccinium myrtillus* L.). *J. Agric. Food Chem.* **2017**, *65*, 7483–7490. [CrossRef]

11. Hornedo-Ortega, R.; Álvarez-Fernández, M.A.; Cerezo, A.B.; Garcia-Garcia, I.; Troncoso, A.M.; Garcia-Parrilla, M.C. Influence of fermentation process on the anthocyanin composition of wine and vinegar elaborated from strawberry. *J. Food Sci.* **2017**, *82*, 364–372. [CrossRef]

12. Liu, S.; Marsol-Vall, A.; Laaksonen, O.; Kortesniemi, M.; Yang, B. Characterization and quantification of nonanthocyanin phenolic compounds in white and blue bilberry (*Vaccinium myrtillus*) juices and wines using UHPLC-DAD–ESI-QTOF-MS and UHPLC-DAD. *J. Agric. Food Chem.* **2020**, *68*, 7734–7744. [CrossRef]

13. Papadopoulou, C.; Soulti, K.; Roussis, I.G. Potential antimicrobial activity of red and white wine. *Phenolic Extracts against strains of Staphylococcus aureus, Escherichia coli and Candida albicans. Food Technol. Biotechnol.* **2005**, *43*, 41–46.

14. Sohn, H.-Y.; Son, K.H.; Kwon, C.-S.; Kwon, G.-S.; Kang, S.S. Antimicrobial and cytotoxic activity of 18 prenylated flavonoids isolated from medicinal plants: *Morus alba* L., *Morus mongolica* Schneider, *Broussnetia papyrifera* (L.) Vent, *Sophora flavescens* Ait and *Echinosophora koreensis* Nakai. *Phytomedicine* **2004**, *11*, 666–672. [CrossRef] [PubMed]

15. Xiao, Z.T.; Zhu, Q.; Zhang, H.Y. Identifying antibacterial targets of flavonoids by comparative genomics and molecular modeling. *Open J. Genom.* **2014**, *3*. [CrossRef]

16. Lee, Y.L.; Cesario, T.; Wang, Y.; Shanbrom, E.; Thrupp, L. Antibacterial activity of vegetables and juices. *Nutrition* **2003**, *19*, 994–996. [CrossRef] [PubMed]

17. Selma, M.V.; Espin, J.C.; Tomas-Barberan, F.A. Interaction between phenolics and gut microbiota: Role in human health. *J. Agric. Food Chem.* **2009**, *57*, 6485–6501. [CrossRef] [PubMed]

18. Clifford, M.; Scalbert, A. Review: Ellagitannins nature, occurence and dietary burden. *J. Sci. Food Agric.* **2000**, *80*, 1118–1125. [CrossRef]

19. Cushnie, T.; Lamb, A.J. Antimicrobial activity of flavonoids. *Int. J. Antimicrob. Agents* **2005**, *26*, 343–356. [CrossRef] [PubMed]

20. Nohynek, L.J.; Alakomi, H.L.; Kahkonen, M.P.; Heinonen, M.; Helander, I.M.; Oksman-Caldentey, K.M.; Puupponen-Pimiä, R.H. Berry phenolics: Antimicrobial properties and mechanisms of action against severe human pathogens. *Nutr. Cancer* **2006**, *54*, 18–32. [CrossRef]

21. Puupponen-Pimia, R.; Nohynek, L.; Hartmann-Schmidlin, S.; Kahkonen, M.; Heinonen, M.; Maatta-Riihinen, K.; Oksman-Caldentey, K.M. Berry phenolics selectively inhibit the growth of intestinal pathogens. *J. Appl. Microbiol.* **2005**, *98*, 991–1000. [CrossRef] [PubMed]

22. Krisch, J.; Galgoczy, L.; Tolgyesi, M.; Papp, T.; Vagvolgyi, C. Effect of fruit juices and pomace extracts on the growth of Gram-positive and Gram-negative bacteria. *Acta Biol. Szeged.* **2008**, *52*, 267–270.

23. Nile, S.H.; Park, S.W. Edible berries: Bioactive components and their effect on human health. *Nutrition* **2014**, *30*, 134–144. [CrossRef] [PubMed]

24. Morrissey, P.A.; Hill, T.R. Vitamins: Vitamin C. In *Encyclopedia of Dairy Sciences*, 2nd ed.; Elsevier, Academic Press: San Diego, CA, USA, 2011; pp. 667–674.

25. Czyzowska, A. Vitamin C. In *Reference Module in Food Sciences*; Elsevier: Amsterdam, The Netherlands, 2015; pp. 1–8. [CrossRef]

26. Velić, D.; Amidžić Klarić, D.; Velić, N.; Klarić, I.; Petravić Tominac, V.; Mornar, A. Chemical Constituents of Fruit Wines as Descriptors of their Nutritional, Sensorial and Health-Related Properties. In *Descriptive Food Science*; InTechOpen: London, UK, 2018; Chapter 4. [CrossRef]

27. Rajkowska, K.; Otlewska, A.; Kunicka-Styczyńska, A.; Krajewska, A. *Candida albicans* impairments induced by peppermint and clove oils at sub-inhibitory concentrations. *Int. J. Mol. Sci.* **2017**, *18*, 1307. [CrossRef]

28. Dewhirst, F.E.; Chen, T.; Izard, J.; Paster, B.J.; Tanner, A.C.R.; Yu, W.-H.; Lakshmanan, A.; Wade, W.G. The Human Oral Microbiome. *J. Bacteriol.* **2010**, *192*, 5002–5017. [CrossRef]

29. EHOMD Expanded Human Oral Microbiome Database. Available online: http://www.homd.org/?name= HOMD&taxonomy_level=1 (accessed on 20 November 2020).

30. Thiyahuddin, N.M.; Lamping, E.; Rich, A.M.; Cannon, R.D. Yeast Species in the Oral Cavities of Older People: A Comparison between People Living in Their Own Homes and Those in Rest Homes. *J. Fungi* **2019**, *5*, 30. [CrossRef]

31. Simões-Silva, L.; Ferreira, S.; Santos-Araujo, C.; Tabaio, M.; Pestana, M.; Soares-Silva, I.; Sampaio-Maia, B. Oral Colonization of Staphylococcus Species in a Peritoneal Dialysis Population: A Possible Reservoir for PD-Related Infections? *Can. J. Infect. Dis. Med. Microbiol.* **2018**, 5789094. [CrossRef]

32. Bottone, E.J. *Bacillus cereus*, a volatile human pathogen. *Clin. Microbiol. Rev.* **2010**, *23*, 382–398. [CrossRef]

33. Czyżowska, A.; Kucharska, A.Z.; Nowak, A.; Sokół-Łętowska, A.; Motyl, I.; Piórecki, N. Suitability of the probiotic lactic acid bacteria strains as the starter cultures in unripe cornelian cherry (*Cornus mas* L.) fermentation. *J. Food Sci. Technol.* **2017**, *54*, 2936–2946. [CrossRef]

34. Rivero-Pérez, M.D.; González-Sanjosé, M.L.; Muñiz, P.; Pérez-Magariño, S. Antioxidant profile of red-single variety wines microoxygenated before malolactic fermentation. *Food Chem.* **2008**, *111*, 1004–1011. [CrossRef]

35. Fogliano, V.; Verde, V.; Randazzo, G.; Ritieni, A. Method for measuring antioxidant activity and its application to monitoring antioxidant capacity of wines. *J. Agric. Food Chem.* **1999**, *47*, 1035–1040. [CrossRef]

36. Efenberger-Szmechtyk, M.; Nowak, A.; Czyżowska, A.; Kucharska, A.Z.; Fecka, I. Composition and Antibacterial Activity of *Aronia melanocarpa* (Michx.) Elliot, *Cornus mas* L. and *Chaenomeles superba* Lindl. Leaf Extracts. *Molecules* **2020**, *25*, 2011. [CrossRef]

37. Duarte, W.F.; Dias, D.R.; Oliveira, J.M.; Vilanova, M.; Teixeira, J.A.; Silva, J.B.D.A.E.; Schwan, R.F. Raspberry (*Rubus idaeus* L.) wine: Yeast selection, sensory evaluation and instrumental analysis of volatile and other compounds. *Food Res. Int.* **2010**, *43*, 2303–2314. [CrossRef]

38. Duarte, W.F.; Dragone, G.; Dias, D.R.; Oliveira, J.M.; Teixeira, J.A.; Silva, J.B.D.A.E.; Schwan, R.F. Fermentative behaviour of *Saccharomyces* strains during microvinification of raspberry juice (*Rubus idaeus* L.). *Int. J. Food Microbiol.* **2010**, *143*, 173–182. [CrossRef] [PubMed]

39. Veberic, R.; Jakopic, J.; Stampar, F.; Schmitzer, V. European elderberry (*Sambucus nigra* L.) rich in sugars, organic acids, anthocyanins and selected polyphenols. *Food Chem.* **2009**, *114*, 511–515. [CrossRef]

40. Johnson, M.H.; Mejia, E.G. Comparison of chemical composition and antioxidant capacity of commercially available blueberry and blackberry wines in Illinois. *J. Food Sci.* **2012**, *77*, C141–C148. [CrossRef]

41. Kalkan Yildirim, H. Evaluation of colour parameters and antioxidant activities of fruit wines. *Int. J. Food Sci. Nutr.* **2006**, *57*, 47–63. [CrossRef]

42. Ortiz, J.; Marin-Arroyo, M.-R.; Noriega-Dominguez, M.-J.; Navarro, M.; Arozarena, I. Color, phenolics, and antioxidant activity of blackberry (*Rubus glaucus* Benth.), blueberry (*Vaccinium floribundum* Kunth.), and apple wines from Ecuador. *J. Food Sci.* **2013**, *78*, C985–C993. [CrossRef]

43. Ljevar, A.; Curko, N.; Tomasevic, M.; Radosevic, K.; Gaurina Srcek, V.; Kovacevic Ganic, K. Phenolic composition, antioxidant capacity and in vitro cytotoxicity of fruit wines. *Food Technol. Botechnol.* **2016**, *54*, 145–155. [CrossRef]

44. Mudnic, I.; Budimir, D.; Modun, D.; Gunjaca, G.; Generalic, I.; Skroza, D.; Katalinic, V.; Ljubenkov, I.; Boban, M. Antioxidant and vasodilatory effects of blackberry and grape wines. *J. Med. Food* **2012**, *15*, 315–321. [CrossRef]

45. Mitic, M.N.; Obradovic, M.V.; Mitic, S.S.; Pavlovic, A.N.; Pavlovic, J.L.J.; Stojanovic, B.T. Free radical scavenging activity and phenolic profile of selected Serbian red fruit wines. *Rev. Chim.* **2013**, *64*, 68–73.

46. Arozarena, I.; Ortiz, J.; Hermosín-Gutiérrez, I.; Urretavizcaya, I.; Salvatierra, S.; Córdova, I.; Marín-Arroyo, M.R.; Noriega, M.J.; Navarro, M. Color, ellagitannins, anthocyanins, and antioxidant activity of andean blackberry (*Rubus glaucus* Benth.) wines. *J. Agric. Food Chem.* **2012**, *60*, 7463–7473. [CrossRef]
47. Gao, X.; Björk, L.; Trajkovski, V.; Uggla, M. Evaluation of antioxidant activities of rosehip ethanol extracts in different test systems. *J. Sci. Food Agric.* **2000**, *80*, 2021–2027. [CrossRef]
48. Brand-Williams, W.; Cuvelier, M.E.; Berset, C. Use of a free radical method to evaluate antioxidant activity. *LWT-Food Sci. Technol.* **1995**, *28*, 25–30. [CrossRef]
49. Lingua, M.S.; Fabani, M.P.; Wunderlin, D.A.; Baroni, M.V. From grape to wine: Changes in phenolic composition and its influence on antioxidant activity. *Food Chem.* **2016**, *208*, 228–238. [CrossRef]
50. Rupasinghe, H.V.; Joshi, V.; Smith, A.; Parmar, I. Chapter 3-Chemistry of Fruit Wines. In *Panesar, Science and Technology of Fruit Wine Production*; Maria, R., Kosseva, V.K., Joshi, P.S., Eds.; Academic Press: Cambridge, MA, USA, 2017; pp. 105–176. ISBN 9780128008508. [CrossRef]
51. Ginjom, I.; D'Arcy, B.; Caffin, N.; Gidley, M. Phenolic compound profiles in selected Queensland red wines at all stages of wine-making process. *Food Chem.* **2011**, *125*, 823–834. [CrossRef]
52. Cerezo, A.B.; Cuevas, E.; Winterhalter, P.; Garcia-Parrilla, M.C.; Troncoso, A.M. Isolation, identification, and antioxidant activity of anthocyanin compounds in *Camarosa* strawberry. *Food Chem.* **2010**, *123*, 574–582. [CrossRef]
53. Hornedo-Ortega, R.; Krisa, S.; Garcia-Parrilla, M.C.; Richard, T. Effect of gluconic and alcoholic fermentation on anthocyanin composition and antioxidant activity of beverages made from strawberry. *LWT-Food Sci. Technol.* **2016**, *69*, 382–389. [CrossRef]
54. Olejnik, A.; Olkowicz, M.; Kowalska, K.; Rychlik, J.; Dembczyński, R.; Myszka, K.; Juzwa, W.; Białas, W.; Moyer, M.P. Gastrointestinal digested *Sambucus nigra* L. fruit extract protects in vitro cultured human colon cells against oxidative stress. *Food Chem.* **2016**, *197*, 648–657. [CrossRef]
55. Sellappan, S.; Akoh, C.C.; Krewer, G. Phenolic compounds and antioxidant capacity of Georgia grown blueberries and blackberries. *J. Agric. Food Chem.* **2002**, *50*, 2432–2438. [CrossRef]
56. Wu, X.; Prior, R.L. Systematic identification and characterization of anthocyanins by HPLC-ESI-MS/MS in common food in the United States: Fruits and berries. *J. Agric. Food Chem.* **2005**, *53*, 2589–2599. [CrossRef]
57. Fan-Chiang, H.J.; Wrolstad, R.E. Anthocyanin pigment composition of blackberries. *J. Food Sci.* **2005**, *70*, C198–C202. [CrossRef]
58. Bowen-Forbes, C.S.; Zhang, Y.; Nair, M.G. Anthocyanin content, antioxidant, antiinflammatory and anticancer properties of blackberry and raspberry fruits. *J. Food Compos. Anal.* **2010**, *23*, 554–560. [CrossRef]
59. Klopotek, Y.; Otto, K.; Böhm, V. Processing strawberries to different products alters contents of vitamin C, total phenolics, total anthocyanins, and antioxidant capacity. *J. Agric. Food Chem.* **2005**, *53*, 5640–5646. [CrossRef]
60. Ubeda, C.; Callejón, R.M.; Hidalgo, C.; Torija, M.J.; Troncoso, A.M.; Morales, M.L. Employment of different processes for the production of strawberry vinegars: Effects on antioxidant activity, total phenols and monomeric anthocyanins. *LWT-Food Sci. Technol.* **2013**, *52*, 139–145. [CrossRef]
61. Buchert, J.; Koponen, J.M.; Suutarinen, M.; Mustranta, A.; Lille, M.; Törrönen, R.; Poutanen, K. Effect of enzyme-aided pressing on anthocyanin yield and profiles in bilberry and blackcurrant juices. *J. Sci. Food Agric.* **2005**, *85*, 2548–2556. [CrossRef]
62. Nowicka, A.; Kucharska, A.Z.; Sokół-Łętowska, A.; Fecka, I. Comparison of polyphenol content and antioxidant capacity of strawberry fruit from 90 cultivars of *Fragaria × ananassa* Duch. *Food Chem.* **2019**, *270*, 32–46. [CrossRef] [PubMed]
63. Kylli, P.; Nohynek, L.; Puupponen-Pimia, R.; Westerlund-Wikstrom, B.; Leppanen, T.; Welling, J.; Moilanen, E.; Heinonen, M. Lingonberry (*Vaccinium vitis-idaea*) and European Cranberry (*Vaccinium microcarpon*) Proanthocyanidins: Isolation, Identification, and Bioactivities. *J. Agric. Food Chem.* **2011**, *59*, 3373–3384. [CrossRef]

64. Puupponen-Pimia, R.; Nohynek, L.; Meier, C.; Kahkonen, M.; Heinonen, M.; Hopia, A.; Oksman-Caldentey, K.M. Antimicrobial properties of phenolic compounds from berries. *J. Appl. Microbiol.* **2001**, *90*, 494–507. [CrossRef]

foods

MDPI

Article

Bioactive Compounds and Antioxidant Capacity in Pearling Fractions of Hulled, Partially Hull-Less and Hull-Less Food Barley Genotypes

Mariona Martínez-Subirà, María-Paz Romero, Alba Macià, Eva Puig, Ignacio Romagosa and Marian Moralejo *

AGROTECNIO-CERCA Center, University of Lleida, Av. Rovira Roure 191, 25198 Lleida, Spain; mariona.martinez@udl.cat (M.M.-S.); mariapaz.romero@udl.cat (M.-P.R.); alba.macia@udl.cat (A.M.); eva.puig@udl.cat (E.P.); ignacio.romagosa@udl.cat (I.R.)
* Correspondence: marian.moralejo@udl.cat; Tel.: +34-97370-2858

Abstract: Three food barley genotypes differing in the presence or absence of husks were sequentially pearled and their fractions analyzed for ash, proteins, bioactive compounds and antioxidant capacity in order to identify potential functional food ingredients. Husks were high in ash, arabinoxylans, procyanidin B3, prodelphinidin B4 and *p*-coumaric, ferulic and diferulic bound acids, resulting in a high antioxidant capacity. The outermost layers provided a similar content of those bioactive compounds and antioxidant capacity that were high in husks, and also an elevated content of tocols, representing the most valuable source of bioactive compounds. Intermediate layers provided high protein content, β-glucans, tocopherols and such phenolic compounds as catechins and bound hydroxybenzoic acid. The endosperm had very high β-glucan content and relative high levels of catechins and hydroxybenzoic acid. Based on the spatial distribution of the bioactive compounds, the outermost 30% pearling fractions seem the best option to exploit the antioxidant capacity of barley to the full, whereas pearled grains supply β-glucans enriched flours. Current regulations require elimination of inedible husks from human foods. However, due to their high content in bioactive compounds and antioxidant capacity, they should be considered as a valuable material, at least for animal feeds.

Keywords: whole barley flour; pearling fractions; proteins; β-glucans; arabinoxylans; tocols; phenolic compounds; antioxidant capacity; functional food

Citation: Martínez-Subirà, M.; Romero, M.-P.; Macià, A.; Puig, E.; Romagosa, I.; Moralejo, M. Bioactive Compounds and Antioxidant Capacity in Pearling Fractions of Hulled, Partially Hull-Less and Hull-Less Food Barley Genotypes. *Foods* **2021**, *10*, 565. https://doi.org/10.3390/foods10030565

Academic Editors: Marina Russo and Francesco Cacciola

Received: 3 February 2021
Accepted: 4 March 2021
Published: 9 March 2021

Publisher's Note: MDPI stays neutral with regard to jurisdictional claims in published maps and institutional affiliations.

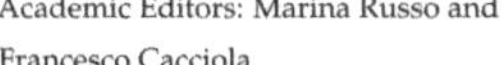

1. Introduction

Barley (*Hordeum vulgare* L.) is the fourth most cultivated cereal in the world after maize, wheat and rice [1]. It is widely used for feed and malt, with limited consumption as human food in some specific regions, such as the Maghreb and the high plateaus of the Himalayas. However, in recent years, it has attracted growing interest worldwide due to the health promoting properties of its bioactive compounds. In terms of health, several reports have demonstrated the positive effect of barley on the glycemic index, cholesterol and heart diseases [2]. This is mainly due to the presence of β-glucans (2–11% d.w.), a dietary fiber component for which health claims have been approved by the US Food and Drug Administration and the European Food Safety Authority [3,4]. Additional beneficial effects have been described for β-glucans. These include their properties as enhancers of the immune system against infectious diseases and some types of cancer, and also as a key modulator of the composition and activity of human microbiota [5–7]. Arabinoxylans are the second major component of barley dietary fiber (2–9% d.w.). Their proven health benefits include effects on the postprandial glucose response, cholesterol metabolism and immune response [8,9]. In barley, arabinoxylans are particularly associated with antioxidant activity due to the presence of phenolic acids linked to their structure [10]. Barley is also a significant source of antioxidant compounds [11,12]. It contains more vitamin E than most cereals (17–49 µg/g d.w.), this being in the form of α-tocopherol and α-tocotrienol. Furthermore, barley contains high levels of phenolic compounds such as phenolic acids and flavan 3-ols which can be found free or ester-bound to the fiber. Flavan 3-ols like catechins, procyanidins and prodelphinidins are the most abundant free phenols, while phenolic acids such as ferulic, *p*-coumaric and vanillic are the major constituents within the bounds. Anthocyanins are also present in considerable concentrations in some barley genotypes with colored kernels. All these phenolic compounds are considered potent antioxidants, free radical scavengers and inhibitors of lipid peroxidation [13,14]. In addition, preclinical studies and clinical trials have shown that polyphenols could greatly modulate the gut microbiota, thus favoring the growth of potential beneficial organisms and simultaneously inhibiting pathogenic bacteria [15].

Barley grains may differ in important morphological characters. These include being hulled (covered), when the husks adhere to the caryopsis, partially hull-less (skinned), when a partial loss of the husks occurs, and hull-less (naked), when the husks are freely threshed at harvest, the latter being what is preferred for human consumption; grains from two or six rowed spikes or grains with colors such black, blue or purple, alternative to yellow. Based on the grain composition, barley is further classified as normal, high amylopectin (waxy) or high amylose starch types, high β-glucan, high lysine, and proanthocyanidin-free. All these types of grain differ widely in their physical and compositional characteristics and, accordingly, are processed differently and used for different commercial purposes.

The bioactive compounds are not uniformly distributed across the barley grains. It is known that arabinoxylans, tocotrienols and phenolics are mainly located in the outer layers, tocopherols in the germ, and β-glucans in the endosperm [16]. Thus, physical processes like pearling, an abrasive technique that gradually removes grain layers to obtain polished grain and by-products, allow favorable separation of fractions enriched in specific compounds and these can be used as functional ingredients. Several fold enrichment of antioxidant compounds has been described in barley pearled fractions [17,18] that have been used to improve the nutritional value of such wheat-based products as cookies, pasta and bread [19–22].

In recent years, increased efforts have been carried in a few countries to release new specific barley varieties for human consumption and for the food industry. In our food barley breeding program, we aim to produce varieties rich in β-glucans as well as antioxidant compounds adapted to the Spanish agro-climatic conditions. Three distinct high β-glucan barley genotypes from our program, differing in the type of grain (Kamalamai, hulled; Hindukusch, partially hull-less; Annapurna, hull-less) were selected to identify specific potential ingredients for the functional food industry. Thus, the main objectives of this work were as follows: (1) to analyze β-glucans, arabinoxylans, tocols, phenolic compounds and antioxidant capacity in the whole flours and pearling fractions of different types of barley grains; (2) to identify pearling fractions to be used potentially as functional ingredients. This research could provide further knowledge about the spatial distribution of a large number of bioactive compounds in barley pearling fractions, since most published works have focused on changes in various bioactive compounds separately [16,17,23]. To the best of our knowledge, this is the first time that such an integrative study of the major health-promoting components in barley genotypes specifically bred for human food has been explored.

2. Materials and Methods

2.1. Plant Material

- Kamalamai: registered Spanish variety, hulled, two rowed, normal endosperm (semillas Batlle SA).
- Hindukusch: Afghan landrace, naked but often suffering from grain skinning, two rowed, normal endosperm and purple grain used as parent in our crosses.

- Annapurna: registered Spanish variety, hull-less, two rowed, waxy endosperm (semillas Batlle SA).

The three genotypes were cultivated under similar conditions in Bell-lloc d'Urgell, Lleida (Spain) during the 2018–2019 growing season.

2.2. Whole Flours and Pearling Fractions

Grain with size above 2.5 mm was screened for this study using a stainless-steel mesh. Six pearling fractions from each genotype were obtained by sequential processing of grain using theTM-05C pearling machine (Satake Corporation, Hiroshima, Japan) at 1060 rpm. The grains were initially pearled to remove the 5% of the original grain weight that resulted in the first fraction F1 (0–5% *w/w*). The remained grains were pearled to remove the second fraction F2 (5–10%), and then the process was repeated to get fractions F3 (10–15%), F4 (15–20%), F5 (20–25%), F6 (25–30%), and the residual 70% pearled grain F7 (30–100% *w/w*). After each pearling session, the pearling machine was cleaned to avoid mixtures between fractions. Fractions, pearled grains and whole grains were ground in a Foss Cyclotec 1093™ mill equipped with a 0.5-mm screen (Foss Iberia, Barcelona, Spain) prior to chemical analyses.

2.3. Protein and Ash Content

The protein content was done according to the Kjeldahl method in a Kjeltec system I (Foss Tecator AB, Höganäs, Sweden) using the conversion factor of 5.7. The ash content was determined in a muffle furnace according to the AOAC Official Method 942.05 [24].

2.4. β-Glucan and Arabinoxylan Content

The β-glucan and arabinoxylan contents were determined by means of the mixed-linkage β-glucan assay (K-BGLU) and D-xylose assay (K-XYLOSE) kits from Megazyme (Wicklow, Ireland).

2.5. Tocols Content

Tocopherols and tocotrienols (α-, β-, γ-, and δ-isomers) were quantified by high performance liquid chromatography (HPLC) coupled to a fluorescence detector. One gram of each barley genotype was extracted three times with 10 mL n-hexane and the extract collected after centrifuging at $9000\times g$ for 10 min. The supernatants were pooled, reduced to dryness under a flow of nitrogen, and reconstituted in 1 mL of n-hexane. Normal phase HPLC with fluorescence detection (excitation 292 nm, emission 325 nm) was used to analyze tocopherols and tocotrienols. Aliquots of 50 µL were injected into the HPLC system following the chromatographic conditions described by Martínez-Subirà et al. [12]. Tocopherols and tocotrienols isomers were quantified with external standard curves. Results were expressed as µg/g dry sample.

2.6. Phenolic Compounds (PCs) and Anthocyanin Analysis by UPLC-MS/MS

Free phenolic compounds were extracted three times by adding 1 mL of 79.5% methanol, 19.5% Milli Q water, and 1% formic acid solution to 150 mg of barley flours. The samples were sonicated for 30 s and centrifuged at $9000\times g$ for 10 min. The supernatants from each extraction were pooled and filtered through 0.22 µm polyvinylidene fluoride (PVDF) filter discs before chromatographic analysis. The residue was subjected to alkaline hydrolysis by adding 6 mL of 2 mol/L NaOH to obtain bound phenols. The samples were left over-night at room temperature for complete hydrolysis. Then, they were sonicated for 1 min and centrifuged at $9000\times g$ for 10 min; the supernatant was acidified with HCl 37% (*w/w*) to pH 2. A total of 350 µL of supernatant was mixed with phosphoric acid 10 min, centrifuged at $9000\times g$, and subjected to µSPE clean-up according to Serra et al. [25]. Briefly, the micro-cartridges were pre-conditioned with acidified water (pH 2) and methanol. The samples were loaded onto the µSPE and subsequently washed with water and water/methanol 95/5 (*v/v*). The PCs were eluted with methanol, and 2.5 µL of the eluate was directly analyzed by liquid chromatography. The extracts were analyzed by Ac Quity Ultra-Performance ^{TM} liquid chromatography coupled to a tandem mass spectrometer (UPLC-MS/MS), equipped with the analytical column Ac Quity BEH C18 (100 mm × 2.1 mm i.d., 1.7 μm) and the Van Guard TM Pre-Column Ac Quity BEH C 18 (5 mm × 2.1 mm, 1.7 μm), all from Waters, Milford, MA, USA. The mobile phase was 0.2% (*v/v*) acetic acid and acetonitrile for the phenolic compounds (PCs), and 10% acetic acid (*v/v*) and acetonitrile for the anthocyanins. The UPLC system was coupled to a triple quadrupole detector mass spectrometer from Waters equipped with a Z-spray electrospray interface for ionization, operating in the negative mode $[M-H]^-$ for the PCs and the positive mode $[M-H]^+$ for the anthocyanins. Quantification was based on a 0.02–25 ng calibration curve of commercially available standards, and the results were expressed as μg/g dry sample. A linear response was obtained for all standards and tested by linear regression analysis. The limits of detection (LOD) ranged from 0.007 to 0.09 ng and the limits of quantification (LOQ) from 0.02 to 0.30 ng.

2.7. Antioxidant Capacity (AC) Analysis

The Oxygen Radical Absorbance Capacity (ORAC) of both the free and bound extracts was measured as described Huang et al. [26]. The antioxidant capacity (AC) was determined using the FLUO star OPTIMA fluorescence reader (BMG Labtech, Offenburg, Germany) in a 96-well polystyrene microplate controlled by the OPTIMA 2.10 R2 software. Changes in fluorescence were measured under controlled temperature (37 °C) in a reader with fluorescence filters with 485 nm excitation and 520 nm emission wavelengths. Trolox (6-hydroxy-2, 5, 7, 8-tetramethylchroman-2-carboxylic acid) was used as control, with one ORAC unit being equal to the antioxidant protection given by 1 μmol Trolox. The results were expressed as μmols of Trolox-equivalents per 100 g of dry sample.

2.8. Statistical Analysis

All statistical analyses were conducted using JMP®Pro14 (SAS institute Inc., Cary, NC, USA). Analytical measurements were carried out in triplicate and the results were presented as mean values. Data was checked for normality and for homoscedasticity of variances based on a number of diagnostic tools, such as residual plots, Box-Cox's Lambda and Levene's test of equality of variances, provided by JMP. When needed, a logarithmic transformation was used and indicated in the text. For multiple comparisons, Tukey–Kramer's honestly-significant-difference tests (HSD) ($\alpha = 0.05$) were conducted once the corresponding ANOVA F-tests were found significant. Principal Component Analysis (PCA) was used to graphically represent the association between fractions, genotypes and bioactive compounds using standardized data.

3. Results and Discussion

3.1. Ash and Protein Contents

The ash and protein contents of the whole flours and pearling fractions are shown in Table 1. The hulled genotype Kamalamai showed the highest ash content and Hindukush, the partially hull-less genotype, the highest protein amount. The distribution through the grain showed a progressive decrease in the ash content from the first pearling fraction F1 of the three genotypes toward the endosperm. This was because the mineral components are mainly localized in the outer layers of the kernel. The protein content was the lowest in the initial surface F1 fraction of the Kamalamai, which mainly correspond to husks, whereas the highest content was detected in the middle fraction F4 of both genotypes containing husks, and F3 of the hull-less Annapurna genotype. These results were in accordance to that observed by other authors on barley fractions obtained by pearling or roller-milling processes [16,22].

Table 1. Ash and protein contents in whole barley flours and pearling fractions.

	Ash (%)			Protein (%)		
	Kamalamai	Hindukush	Annapurna	Kamalamai	Hindukush	Annapurna
Whole flour	1.97 a	1.68 a	1.48 b	14.62 b	15.26 a	14.83 ab
Fractions						
F1	6.64 a	6.52 a	5.16 a	7.77 g	10.54 e	19.97 e
F2	5.60 b	5.75 b	4.74 b	17.88 e	17.17 d	25.63 c
F3	5.00 c	5.86 b	3.81 c	24.18 d	23.10 c	27.88 a
F4	3.93 d	5.09 c	3.23 d	26.72 a	27.57 a	27.34 b
F5	3.17 e	3.94 d	2.26 e	26.11 b	27.34 a	25.33 c
F6	2.50 f	2.76 e	1.77 f	25.21 c	26.81 b	23.11 d
F7	0.87 g	0.76 f	1.13 g	12.80 f	10.54 f	11.93 f
SED	0.04	0.05	0.03	0.13	0.11	0.14

Results are presented as the mean. Means within a column followed by different letters indicate significant differences; (Tukey-Kramer's HSD for α = 0.05) SED: standard error of the difference between means. F1–7: fraction1–7.

3.2. β-Glucans, Arabinoxylans and Tocols Contents

In barley, the β-glucan content depends on genetic and environmental factors as well as on the interaction between these [8,27]. In our work, the three barley genotypes were high in β-glucans with amounts ranging from 8.3 to 9.5 g/100 g (Table 2). Annapurna had the highest β-glucan content while both genotypes containing husks showed similar values. These results agree with earlier studies which described higher β-glucan levels in hull-less and waxy genotypes like Annapurna rather than in hulled barley with normal endosperm [2,28]. The distribution of this fiber component through the grain is shown in Figure 1. β-glucans increased gradually from the outer to inner layers of the grains in accordance with previous findings [22,29]. The lowest concentrations were detected in the outer fraction F1 of the three genotypes and progressively increased until F4 of Annapurna, F5 of Kamalamai, and F6 of Hindukusch, after wich they remained constant. The net effect was that removing the 30% outer fractions of Kamalamai and Hindukusch increased the β-glucan content by 9% and 6%, respectively, while this increase was only 3% in the hull-less genotype Annapurna.

Table 2. β-glucans, arabinoxylans and tocols contents in whole barley flours.

	Kamalamai	Hindukusch	Annapurna
β-glucans (g/100 g)	8.41 ± 0.09 b	8.26 ± 0.07 b	9.46 ± 0.10 a
Arabinoxylans (g/100 g)	5.52 ± 0.27 b	6.60 ± 0.44 a	5.52 ± 0.19 b
α-Tocopherol (µg/g)	8.32 ± 0.21 a	7.53 ± 0.22 a	6.22 ± 0.19 b
β-Tocopherol (µg/g)	0.03 ± 0.01 a	0.03 ± 0.01 a	0.03 ± 0.01 a
γ-Tocopherol (µg/g)	1.58 ± 0.00 a	1.08 ± 0.01 c	1.21 ± 0.01 b
δ-Tocopherol (µg/g)	0.03 ± 0.00 a	0.02 ± 0.00 c	0.03 ± 0.00 b
Total Tocopherol (µg/g)	9.97 ± 0.05 a	8.66 ± 0.06 b	7.49 ± 0.05 c
α-Tocotrienol (µg/g)	17.03 ± 0.42 b	20.06 ± 0.90 a	18.84 ± 0.23 a,b
β-Tocotrienol (µg/g)	2.05 ± 0.00 b	3.93 ± 0.15 a	3.61 ± 0.04 a
γ-Tocotrienol (µg/g)	9.06 ± 0.20 a	9.05 ± 0.46 a	5.63 ± 0.03 b
δ-Tocotrienol (µg/g)	2.07 ± 0.02 b	2.72 ± 0.07 a	2.19 ± 0.03 b
Total Tocotrienol (µg/g)	30.21 ± 0.16 b	35.77 ± 0.40 a	30.27 ± 0.08 b
Total Tocols (µg/g)	40.18 ± 0.81 a,b	44.43 ± 0.82 a	37.77 ± 0.81 b

Results are presented as the mean ± standard error of the mean. Means within rows followed by different letters indicate significant differences; (Tukey-Kramer's HSD for α = 0.05).

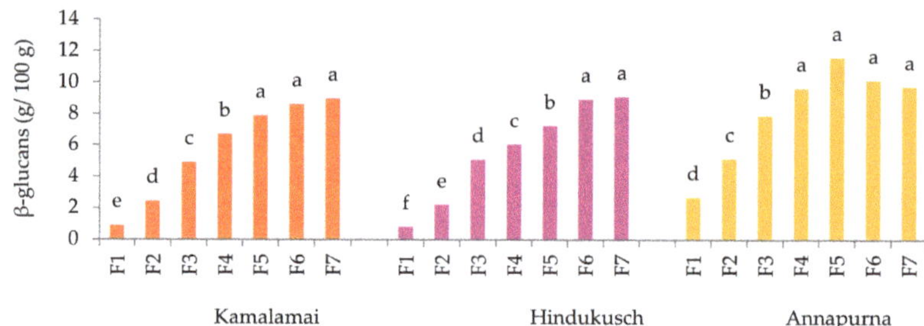

Figure 1. β-Glucan content (g/100 g) in barley pearling fractions. The results are presented as the mean; different letters indicate significant differences within the pearled fraction of each genotype; (Tukey-Kramer's HSD for α = 0.05). F1–7: fraction1–7.

Arabinoxylans in cereals are mainly localized in the husks and cell walls of the outer layers of the grain including pericarp, testa and aleurone. In barley, aleurone cell walls are built up mainly of arabinoxylans (60 to 70%) whereas the endosperm cell walls contain only about 20 to 40% [30]. In our work, the arabinoxylan content of whole barley flours ranged from 5.5 to 6.6 g/100 g; Hindukusch being the highest (Table 2). These values were in accordance with those reported for different barley accessions in a previous work [31]. Contrary to the β-glucans, the arabinoxylans decreased progressively from the outer to the inner layers of the grain (Figure 2). The highest arabinoxylans amounts were observed in all F1 fractions. These mainly correspond to the hulls, testa and to the pericarp of both genotypes containing husks, and to pericarp, testa and some aleurone layers of Annapurna. The arabinoxylan level detected in the outermost layer F1 of the three genotypes was on average four times higher than that in whole barley flours. This finding may draw attention to barley husks as a good source of arabinoxylans.

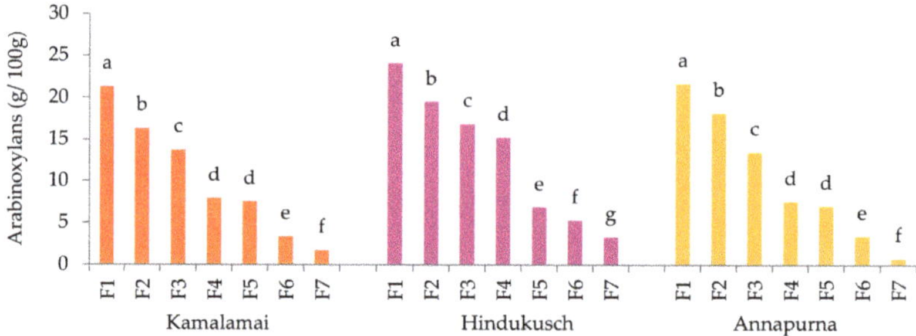

Figure 2. Arabinoxylan contents (g/100 g) in barley pearling fractions. Results are presented as the mean; different letters indicate significant differences within pearled fraction of each genotype; (Tukey-Kramer's HSD for α = 0.05).

Tocols are plant metabolites with interest for their potential benefits for human health [32]. Tocols, which consist of tocopherol and tocotrienols, are found in cereals at moderate levels ranging between 17 and 49 μg/g [17,33]. Among cereals, barley is one of the best sources of tocols due to the high content and favorable distribution of all eight major tocols, α-, β-, γ- and δ- tocopherols and tocotrienols [34]. While all tocol forms have similar antioxidant properties, α tocopherol (αT) is the only one that meets the Recommended Daily Allowance for vitamin E [35,36]. In our study, there were detectable concentrations of the eight tocol isomers in all whole barley flours (Table 2). Total tocols ranged from 38 to 44 μg/g in good accordance with values found in other barley cultivars [12,16–18]. Tocotrienols accounted for 79% of the total tocols while the tocopherols were 21%. αT and α tocotrienols (αT3) were the main isomers in each tocol class, and their concentrations ranged from 6.2 to 8.3 μg/g and 17 to 20 μg/g respectively. Significant differences were identified among genotypes for most tocol forms. Both genotypes containing husks were the highest in αT, γT3 and total tocols. The tocopherol and tocotrienol contents in the fractions are shown in Figure 3A,B and Table S1. In line with previous finding, tocopherols and tocotrienols are distributed in a tissue specific manner with tocopherols mainly located in the germ whit tocotrienols in the aleurone and sub-aleurone layers [16,17,35]. In this study, the highest tocopherol contents were detected between the middle fractions F3 to F6 of genotypes containing husk. This correspond to the highest amount of protein as described above, whereas the content was uniformly distributed across the grain in the hull-less genotype Annapurna. Tocotrienols were more abundant in the outer layers F2 and F3 of the genotypes containing husks, and F1 to F3 of the hull-less one. Fractions F2 to F4 of the genotypes containing husks, and F1 to F3 of Annapurna would provide average tocol contents of 194 μg/g. These selected fractions contain over five times more tocols than whole flours and could be used as a valuable source of natural tocols.

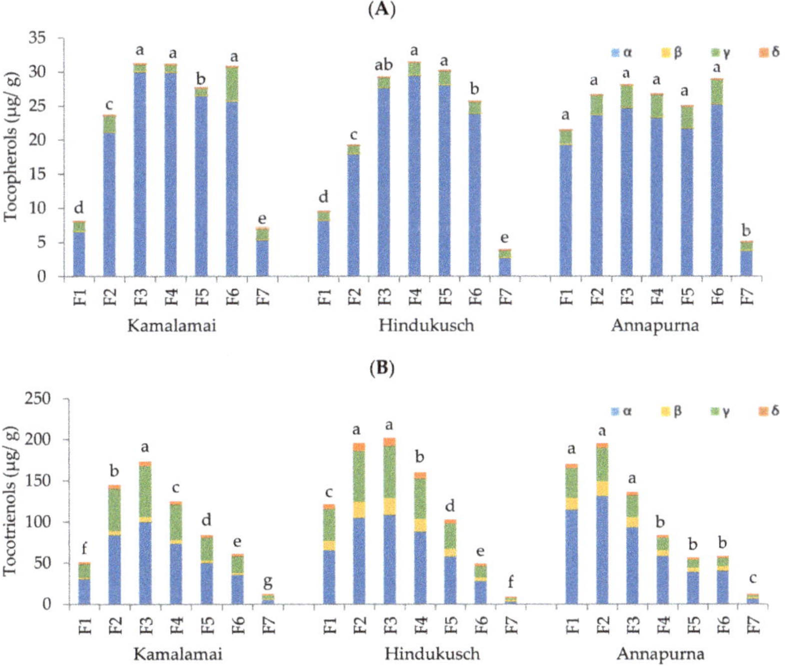

Figure 3. (**A**) α-, β-, γ-, δ- Tocopherol contents (μg/g), and (**B**) α-, β-, γ-, δ- Tocotrienol contents (μg/g) in barley pearling fractions. Results are presented as the mean; different letters indicate significant differences between pearled fractions of each genotype on log-transformed data; (Tukey-Kramer's HSD for α = 0.05).

3.3. Phenolic Compounds (PCs) Contents

Phenolic compounds in barley can be found free or bound to the fiber being differentially distributed across the grain [37]. The analysis of PCs in whole barley flours carried out by ultra-performance liquid chromatography-tandem mass spectrometer (UPLC-MS/MS) included free and bound forms (Table 3). Total free PCs ranged from 369 to 600 μg/g among which, flavan-3-ols accounted for 80%, phenolic acids for 16% and flavone glycosides for 4.1%. Total bound PCs ranged from 781 to 1194 μg/g and were comprised of 99.8% of phenolic acids and 0.2% of flavone glycosides. Bound phenolic acids were predominant in all whole flours representing 65–76% of the total PCs.

Table 3. Phenolic compounds contents (µg/g) in whole barley flours.

	Kamalamai	Hindukusch	Annapurna
Flavan-3-ols	500.7 ± 33.2 a	272.3 ± 12.8 b	320.7 ± 14.6 b
Free phenolic acids	81.1 ± 6.1 a	70.6 ± 2.3 a,b	59.6 ± 2.3 b
Free flavones glycosides	18.4 ± 0.5 b	26.0 ± 0.4 a	9.0 ± 0.2 c
Total free	600.2 ± 30.9 a	368.8 ± 14.8 b	389.3 ± 12.2 b
Bound phenolic acids	1092.1 ± 113.4 a,b	1192.6 ± 215.6 a	779.5 ± 32.9 b
Bound flavones glycosides	1.9 ± 0.1 a	1.8 ± 0.0 a	1.8 ± 0.0 a
Total bound	1093.9 ± 113.5 a,b	1194.4 ± 215.6 a	781.3 ± 32.9 b
Total phenols	1694.1 ± 117.4 a	1563.2 ± 211.5 a,b	1038.6 ± 36.7 b

Results are presented as the mean ± standard error of the mean. Means within rows followed by different letters indicate significant differences; (Tukey-Kramer's HSD for α = 0.05).

Looking at the pearling fractions, the results showed a wide range of phenolic contents with differences between fractions and genotypes. Twenty-two different free phenols were detected. Their distribution within the fractions is detailed in Table 4. Procyanidin B3 and Prodelphinidin B4 were the most abundant flavan-3-ols in all the external fractions. Fractions F2 and F3 of the genotypes containing husks and F1 and F2 of Annapurna showed the highest concentrations of most flavan-3-ols which then progressively decreased until the endosperm except for catechins whose content did not vary as much between fractions. Fraction F1 was the richest in free phenolic acids in the three genotypes. This excludes gallic acid (the major free phenolic acid) whose content was similar within the external fractions. Some free phenolic acids, such as decarboxylated diferulic, hidroxybenzoic, caffeic and cinnamic, were exclusively detected in F1, F2 and F3 and were absent in the rest of the grain. Moreover, decarboxylated diferulic, hydroxybenzoic and 2,4-dihydroxybenzoic were not detected in Annapurna. F1 and F2 had also the highest concentrations of some free flavone glycosides such as apigenin 6-C-arabinoside 8-C-glucoside and ixovitexin 7-rutinoside, while the concentration of the remained free flavones was homogeneous or randomly distributed across the grain. Isoorientin was not detected in Annapurna nor was isovitexin 7-(6‴-sinapoylglucoside) 4′-glucoside in Hindukusch and Annapurna.

With reference to bound phenolic acids, their distribution across the grain was similar in the three genotypes (Table 5). The highest values were detected in the outermost fraction (F1) and gradually decreased toward the core of the grain, as did arabinoxylans, with which they are esterified. In fact, arabinoxylans positively correlated with bound phenolic acids (r = 0.93, p < 0.001). Ferulic acid was found to be the most abundant bound phenolic acid found in the three genotypes, accounting for 68–83% of total bound; p-coumaric acid came second in the genotypes containing husks (12–15%) and the diferulic acid in Annapurna (6%). Several authors detected high amounts of bound phenolic acids in husks [29,30]. This might explain the results observed in our study: the genotypes containing husks being the richest in these compounds. In general, F1 was remarkably high in bound phenolic acids and might provide an important source of antioxidants. Minor bound phenolic acids, such as hydroxybenzoic acid, were uniformly distributed across the grain, and the distribution of others, like isoferulic, hydroxybenzoic, 2,4-dihydroxybenzoic, caffeic, sinapic and cinnamic acids, was not very marked. Bound diferulic tetrahydrofuran acid seemed to be exclusively present in the husks and pericarp layers as it was only detected in the F1 and F2 of both genotypes containing husks. Bound flavone glycosides showed similar distribution pattern as did the free forms.

Table 4. Free phenolic compounds contents in barley pearling fractions.

	Flavan-3-ols (µg/g)				Phenolic Acids (µg/g)											Flavones Glycosides (µg/g)						
	Cat	Cat-g	Pc B3	Pd B4	Gallic	Ferulic	DC diF	p-Cm	p-OHB	OHB	2,4-diOHB	Vanill	Caff	Syring	Cinna	Ap g	Isosc g	Isosc r	Isoor	Isov g	Isov g-g	Isov r
										Kamalamai												
F1	53 d	102 b	374 c	312 c	35.9 a	27.1 a	1.1 a	25.7 a	21.5 a	10.9 a	16.9 a	78.4 a	11.6 b	23.3 a	12.8 a	13.8 a	13.4 a,b	8.3 a	7.3 a	0.2 d	1.7 a	1.8 a
F2	96 c	173 a	662 a	536 b	37.0 a	20.0 a	1.1 a	17.2 a	16.5 b	10.8 a	14.5 b	48.1 b	12.8 a	17.2 b	11.5 b	10.0 b	14.7 a	8.1 a	7.1 a,b	0.2 c,d	1.7 a	1.1 a,b
F3	125 a,b	194 a	740 a	644 a	39.8 a	12.8 b	-	6.6 b	12.8 c	-	-	14.6 c	12.6 a	10.6 c	-	3.4 c	10.4 d,e	7.2 b	6.6 b	0.4 b,c	0.3 b	0.7 b,c
F4	143 a	125 b	434 b	322 c	38.8 a	7.2 c	-	3.5 b,c	11.3 d	-	-	8.3 d	-	8.3 d	-	1.4 d	11.4 c,d	7.1 b	6.7 a,b	0.7 a	-	0.4 c
F5	121 a,b	76 c	264 d	192 d	37.4 a	7.0 c,d	-	3.1 b,c	11.1 d	-	-	8.6 d	-	8.3 d	-	1.0 e	12.8 b,c	7.0 b	6.9 a,b	0.7 a	-	0.4 c
F6	104 b,c	56 d	202 e	134 e	28.3 b	5.3 d,e	-	2.7 c	10.9 d	-	-	7.0 d	-	7.5 d	-	0.7 f	12.3 b,c	6.9 b	7.0 a,b	0.5 a,b	-	0.4 c
F7	73 d	19 e	61 f	36 f	18.0 c	4.1 e	-	0.4 d	1.5 e	-	-	2.5 e	-	1.1 e	-	0.1 g	9.5 e	1.0 c	1.0 c	0.1 e	-	0.1 d
SED	0.1	0.1	0.1	0.1	0.1	0.1	0.1	0.3	0.03	0.1	0.4	0.1	0.2	0.1	0.1	0.1	0.04	0.02	0.03	0.2	0.1	0.2
										Hindukusch												
F1	34 c	53 b	227 b	385 b	24.1 a,b	24.6 a	1.9 a	14.2 a	21.8 a	15.5	40.6 a	58.4 a	11.7 a	15.0 a	-	11.3 a	8.3 b	9.1 a	7.7 a	1.5 a	-	2.2 a
F2	48 a,b	100 a	359 a	612 a	27.8 a,b	14.3 b	1.1 b	5.9 b	18.8 b	-	31.3 b	44.7 b	10.8 b	11.4 b	-	7.2 b	8.4 b	9.1 a	7.4 a,b	2.1 a	-	2.0 a,b
F3	58 a	101 a	349 a	544 a	29.5 a,b	10.2 c	0.7 c	3.7 b,c	16.5 c	-	21.7 c	36.2 c	10.5 b	9.4 c	-	4.3 c	8.3 b	8.0 b	7.0 b,c	1.5 a	-	1.3 b,c
F4	46 b	56 b	235 b	316 c	31.1 a	7.8 d	-	3.0 c,d	15.3 d	-	15.9 d	28.8 d	-	8.4 c,d	-	2.1 d	7.5 c	7.5 b,c	6.9 c	0.8 b	-	0.9 c,d
F5	33 c	24 c	107 c	129 d	32.6 a	6.2 d	-	1.9 d,e	14.1 e	-	14.1 d	23.5 e	-	7.9 d	-	1.2 e	8.7 b	7.3 c	6.9 c	0.6 b,c	-	0.7 d
F6	27 c	14 d	49 d	48 e	20.5 b,c	4.4 e	-	1.1 e	13.1 e	-	1.9 e	17.2 f	-	7.6 d	-	0.6 f	9.5 a	7.1 c	6.9 c	0.3 c	-	0.3 e
F7	16 d	2 d	13 e	8 f	16.1 c	3.5 f	-	-	10.5 f	-	-	5.9 g	-	0.1 e	-	0.1 g	8.3 b	1.0 d	1.0 d	0.1 d	-	0.03 f
SED	0.1	0.1	0.03	0.1	0.1	0.1	0.1	0.2	0.02	0.1	0.1	0.1	0.2	0.04		0.1	0.03	0.02	0.02	0.2		0.2
										Annapurna												
F1	91 b	349 a	927 a	554 a	25.4 a	11.9 a	-	4.5 a	12.4 a	-	-	25.8 a	11.3 a	11.7 a	12.1 a	3.7 a	10.7 a	8.5 a	-	0.2 b,c	-	1.0 a
F2	85 c	245 a	601 b	339 a,b	24.1 a	8.1 b	-	2.2 b	10.9 b	-	-	16.9 a,b	10.3 b	9.2 b	11.4 a,b	1.4 b	8.7 d	7.2 b	-	0.3 a,b	-	0.5 b
F3	91 b,c	139 b	302 c	164 b	22.3 a	6.1 b,c	-	1.8 b,c	10.9 b	-	-	12.5 b	-	7.9 c	10.9 b	0.5 c	9.0 c,d	7.0 b	-	0.3 a,b	-	0.2 c
F4	105 a	79 c	156 d	71 c	21.3 a	4.7 c,d,e	-	1.8 b,c	10.7 b	-	-	6.5 c	-	7.3 c	-	0.4 c,d	9.7 b,c	-	-	0.4 a,b	-	0.2 c
F5	90 b,c	53 c	108 e	49 c	21.0 a	5.2 c,d	-	1.6 c	10.2 b,c	-	-	6.4 c	-	7.1 c	-	0.4 c,d	11.3 a	-	-	0.5 a	-	0.2 c
F6	89 b,c	29 d	55 f	20 d	22.2 a,b	3.6 e	-	-	10.3 b,c	-	-	4.3 c	-	7.1 c	-	0.2 d	10.6 a,b	-	-	0.1 c	-	-
F7	51 d	15 e	30 g	15 d	16.4 b	4.2 d,e	-	-	9.8 c	-	-	2.2 d	-	1.0 d	-	-	7.7 e	-	-	-	-	-
SED	0.1	0.1	0.1	0.2	0.1	0.1		0.1	0.02			0.1	0.1	0.03	0.3	0.2	0.03	0.01		0.2		0.1

Results are presented as the mean. Means within a column followed by different letters indicate significant differences on log-transformed data; (Tukey-Kramer's HSD for $\alpha = 0.05$). SED: standard error of the difference between means. Cat: Catechin, Cat-g: Catechin-glucoside, Pc B3: Procyanidin B3, Pd B4: Prodelphinidin B4, DCDiF: Decarboxylated diferulic acid, p-Cm: p-coumaric acid, p-OHB: p-hydroxybenzoic acid, OHB: m- or o-hydroxybenzoic acid, 2,4-diOHB: 2,4-dihydroxyenzoic acid, Vanill: Vanillic acid, Caff: Caffeic acid, Syring: Syringic acid, Cinna: Cinnamic acid, Ag g: Apigenin 6-C-arabinoside 8-C-glucoside, Isosc g: Isoscoparin 7-glucoside, Isosc r: Isoscoparin 7-rutinoside, Isoor: Isoorientin, Isov g: Isovitexin 7-glucoside, Isov g-g: Isovitexin 7-(6"'-sinapoylglucoside) 4'-glucoside, Isov r: Isovitexin 7-rutinoside.—Not detected.

Table 5. Bound phenolic compounds contents (μg/g) in barley pearling fractions.

	Phenolic Acids (µg/g)																Flavones Glycosides (µg/g)			
	Ferulic	isoF	DiF	TriF	DC DiF	THF DiF	p-Cm	m-Cm	p-OHB	OHB	2,4-diOHB	Vanillic	Caffeic	Syringic	Sinapic	Cinna	Ag g	Isosc-g	Isosc r	Isov g
									Kamalamai											
F1	1367 a	17 b,c	330 a	21.2 a	34.3 a	5.5 a	978 a	21.3 a	22.8 a	3.0 c	3.5 b,c	117 a	5.1 c	19.6 a	14.1 a,b	3.8 d	1.6 a	1.7 d	-	0.05 b
F2	1260 a	15 b,c	165 b	15.8 a	18.2 b	1.9 b	523 b	8.1 b	15.3 b	4.6 a,b	4.2 b	78 b	7.2 b	15.9 a	20.3 a	5.5 a	1.1 a	2.4 a	-	0.09 a
F3	1184 a	15 b,c	151 b	31.7 a	3.0 c	-	182 c	3.5 c	11.1 c	4.9 a,b	6.0 a	44 c	16.1 a	9.4 b	21.1 a	5.9 a	0.2 b	2.1 b	-	-
F4	804 b	20 b	69 c	3.8 b	1.1 d	-	65 d	2.2 c,d	6.8 d	4.8 a,b	2.8 c	18 d	3.9 c,d	4.9 c	11.2 b,c	4.5 b	0.1 c	2.0 b,c	-	-
F5	833 b	32 a	70 c	2.2 b	0.9 d,e	-	53 d,e	3.4 c	6.6 d	5.4 a	-	17 d	3.8 c,d	4.4 c	10.0 b,c	4.4 b,c	0.1 c	2.1 b,c	-	-
F6	575 c	14 c	46 d	2.5 b	0.7 e	-	38 e	1.6 d	5.1 e	4.2 b	-	11 e	2.8 d	3.3 d	7.1 c	3.9 c,d	0.1 c	1.9 c	-	-
F7	206 d	6 d	17 e	1.8 b	0.1 f	-	14 f	0.9 e	3.3 f	4.3 b	-	4 f	0.4 e	2.1 e	1.1 d	3.1 e	0.01 d	0.3 e	-	-
SED	0.1	0.1	0.1	0.3	0.1	0.1	0.1	0.2	0.1	0.3	0.3	0.1	0.1	0.1	0.2	0.04	0.3	0.03		0.01
									Hindukusch											
F1	2030 a	16 b,c	342 a	29.5 a	18.7 a	1.9 a	939 a	12.1 a	45.8 a	4.7 a	37.3 a,b	128 a	5.2 b,c	14.7 a	44.1 a,b	5.3 b	0.8 a	2.0 a	1.7 a	0.2 a
F2	2237 a	20 a,b	374 a	26.2 a	9.2 b	0.6 b	355 b	3.7 b	48.0 a	5.3 a	77.5 a	135 a	6.8 a,b	11.5 a,b	57.4 a	6.3 a	0.3 b	2.0 a,b	1.7 a	0.1 b
F3	1968 a	25 a,b	309 a	23.1 a	5.8 b	-	154 c	1.7 c	33.3 b	5.3 a	63.9 a	99 a,b	8.7 a	10.3 b,c	47.9 a,b	5.6 a,b	0.2 b	1.8 b,c	1.6 a	0.1 b
F4	1643 a	30 a	204 b	10.1 b	3.0 c	-	111 c	1.5 c	22.5 c	5.7 a	26.3 b	69 b	4.4 c	8.4 c	32.1 b	5.0 b	0.1 c	1.7 c	1.6 a	0.1 b
F5	1130 b	24 a,b	114 c	6.7 b,c	1.9 c,d	-	53 d	1.1 c	13.8 d	5.6 a	7.4 c	41 c	3.1 d	5.7 d	17.1 c	4.2 c	0.1 c	1.8 b,c	1.1 a	0.1 b
F6	840 b	25 a,b	74 d	4.2 c	1.3 d	-	35 d	1.3 c	10.1 e	5.3 a	4.2 c	29 c	2.8 d	4.4 d	12.2 c	3.7 c	0.1 c	1.9 a,b,c	1.6 a	0.04 c
F7	353 c	10 c	24 e	1.2 c	0.2 e	-	9 e	0.1 d	5.3 f	5.4 a	3.3 c	9 d	0.4 e	2.4 e	1.1 d	3.2 d	-	0.3 d	-	-
SED	0.1	0.2	0.1	0.2	0.7	0.1	0.2	0.2	0.1	0.4	0.2	0.1	0.1	0.1	0.2	0.04	0.2	0.03	0.01	0.2
									Annapurna											
F1	1850 a	15 b,c	262 a	28.5 a	5.2 a	0.30	169 a	1.7 a	11.1 a	4.6 a,b	-	43 a	5.9 a,b	10.2 a	42.0 a	6.3 a	0.2 a	2.3 a	-	-
F2	1697 a	16 b,c	179 b	13.3 b	1.6 b	-	78 b	0.8 b	8.7 b	4.9 a	-	26 b	11.7 a	7.2 b	27.9 b	5.1 b	0.1 b	2.2 a,b	-	-
F3	1323 b	19 a,b	109 c	3.8 c	0.9 c	-	37 c	0.6 b,c	6.7 c	4.6 a,b	-	16 c	3.6 b	5.2 c	15.4 c	4.4 b	-	2.1 b,c	-	-
F4	968 c	25 a	69 d	3.1 c,d	1.0 c	-	21 d	0.5 b,c	4.9 d	4.3 b,c	-	9 d	2.8 b	3.5 d	8.3 d	3.5 c	-	2.0 c,d	-	-
F5	809 d	20 a,b	58 d	2.1 d,e	2.3 b	-	20 d	0.5 b,c	4.1 e	4.0 c	-	7 d,e	2.7 b	2.8 e	5.4 e	3.4 c,d	-	1.9 d	-	-
F6	623 e	14 c	42 e	2.3 d	2.1 b	-	15 d	0.5 b,c	3.9 e	4.3 b,c	-	6 e	2.6 b	2.5 e	4.1 e	3.3 c,d	-	1.9 d	-	-
F7	259 f	6 d	18 f	1.4 e	0.1 d	-	7 e	0.4 c	3.1 f	4.1 c	-	3 f	-	1.9 f	-	2.9 d	-	-	-	-
SED	0.1	0.1	0.1	0.1	0.1	0.0	0.1	0.2	0.04	0.1		0.1	0.3	0.1	0.1	0.1	0.1	0.02		

Results are presented as the mean. Means within a column followed by different letters indicate significant differences on log-transformed data; (Tukey-Kramer's HSD for α = 0.05). SED: standard error of the difference between means. iso-F: isoferulic, DiF: Diferulic acid, TriF: Triferulic acid, DC DiF: Decarboxylated diferulic acid, DiF THF: Diferulic tetrahydrofuran, *p*-Cm: *p*-Coumaric acid, *m*-Cm: *m*-Coumaric acid, *p*-OHB: *p*–hydroxybenzoic acid, OHB: *m*- or *o*-hydroxybenzoic acid, 2,4DiOHB: 2,4-dihydroxyenzoic acid, Cinna: Cinnamic acid, Ag-g: Apigenin 6-C-arabinoside 8-C-glucoside, Isosc g: Isoscoparin 7-glucoside, Isosc r: Isoscoparin 7-rutinoside, Isov g: Isovitexin 7-glucoside.—Not detected.

Considering total PCs, similar distribution patterns between pearling fractions were observed in the three genotypes (Figure 4). F1 to F3 of the genotypes containing husks were those with the highest total PCs with the average content of 4339 µg/g, being on average 3 times higher than the contents in whole flours. On the contrary, F1 of the hull-less genotype was the highest in total PCs.

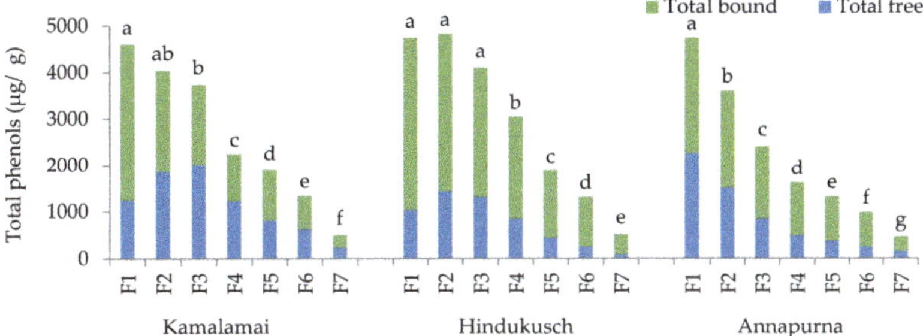

Figure 4. Total phenols contents (µg/g) in seven pearling fractions of three food barley genotypes. Results are presented as the mean; different letters indicate significant differences between pearled fractions of each genotype on log-transformed data; (Tukey-Kramer's HSD for $\alpha = 0.05$).

Most studies into the spatial distribution of PCs in barley have focused on either the total PCs by colorimetric methods [21,22] or just the major phenolic acids [38,39], or have analyzed representative phenolic acids and free flavan-3-ols in a few fractions of previously dehulled cultivars [18,40]. The results obtained in the present study show for the first time the specific distribution of a great variety of phenolic compounds determined by HPLC-MS/MS. These include phenolic acids, flavan-3-ols and flavone glycosides in seven pearling fractions in three food barley genotypes differing in the presence of the husk in the threshed grain.

3.4. Anthocyanins

In this work, the anthocyanin content of Kamalamai and Annapurna was negligible, whereas a total of 24 anthocyanins were detected in the purple Hidukusch genotype. These included pelargonidin, cyanidin, peonidin, delphinidin, petunidin and malvinidin conjugates of glucose, acetylglucose, malonylglucose, dimalonylglucose, arabinose, rutinose and dihexose. The total anthocyanin content was 47 µg/g, which was lower than or similar to those detected in other purple barley cultivars [12,41,42]. This indicates a great diversity, probably associated with genetic and environmental factors. Cyanidin dimalonyl glucoside represented 45% of the total anthocyanins followed by cyaniding glucoside, which accounted for 35%. Several studies have shown that anthocyanins are mainly concentrated in the pericarp and some aleurone layers that provide the grain colour [35]. In this study, the distribution of anthocyanins between the barley fractions showed fractions F1 and F2 as the highest (Table 6 and Table S2). The total anthocyanin contents detected in F2 was upon 10 times higher than that in whole flour. Like most phenolic compounds, the anthocyanin concentration decreased progressively from F2 to F7 where minor contents were detected.

Table 6. Anthocyanins contents (µg/g) in barley pearling fractions.

	Pelagonidins	Cyanidins	Peonidins	Delphinidins	Petunidins	Malvinidins	Total Anthocyanins
				Hindukusch			
F1	41.0 a	285.3 a	5.87 b	17.8 a	3.42 a	1.15 a	354.5 a
F2	50.0 a	366.3 a	21.59 a	20.3 a	2.01 a	1.29 a	461.2 a
F3	25.0 b	188.6 b	0.45 b,c	12.6 b	1.50 a,b	1.06 a	229.2 b
F4	9.6 c	72.5 c	0.25 b,c	6.4 c	0.89 a,b	0.61 a,b	90.2 c
F5	4.6 d	32.8 d	0.17 b,c,d	3.4 d	0.59 b,c	0.40 a,b,c	42.0 d
F6	2.7 e	19.3 e	0.10 c,d	2.1 e	0.20 c	0.30 b,c	24.7 e
F7	1.1 f	7.0 f	0.02 d	0.7 f	0.04 d	0.14 c	9.0 f
SED	0.1	0.1	0.7	0.1	0.3	0.3	0.1

Results are presented as the mean. Means within a column followed by different letters indicate significant differences on log-transformed data; (Tukey-Kramer's HSD for α = 0.05). SED: standard error of the difference between means.

3.5. Antioxidant Capacity (AC)

Antioxidant capacity is an important integrative parameter for evaluating the potential health benefits of foods. The oxygen radical absorbance capacity (ORAC) values of the free, bound and total phenolic compounds of whole barley flours were in the ranges from 46 to 71, 28 to 58, and 74 to 119 µmol Trolox/g respectively (Table 7).

Table 7. Antioxidant Capacity (µmol Trolox/g) of whole barley flours.

	Kamalamai	Hindukusch	Annapurna
Free AC	70.7 ± 1.3 a	61.9 ± 0.2 b	45.6 ± 0.9 c
Bound AC	39.5 ± 2.4 b	57.4 ± 3.3 a	28.1 ± 0.9 c
Total AC	110.2 ± 1.1 a	119.3 ± 3.5 a	74.0 ± 1.8 b

Results are presented as the mean ± standard error of the mean. Means within a rows followed by different letters indicate significant differences; (Tukey-Kramer's HSD for α = 0.05).

The contribution of free PCs to the total AC was higher than that of the bound PCs. This suggested that free PCs had excellent AC as determined by the ORAC assay, since they were at much lower concentrations than bound PCs as explained above. The average ORAC values of genotypes containing husks were significantly higher than that of Annapurna. The ORAC values detected in the three barley genotypes were similar or relatively higher than those reported for whole barley flours in previous works [23,43–45].

When the ORAC assay was measured in the pearling fractions, the highest values were found in the first three fractions of the genotypes containing husks and F1 of Annapurna (Figure 5). F1 of Kamalamai showed a similar AC to F2 and F3 despite having higher amounts of total PCs. These results should be attributed to the strong antioxidant capacity of free PCs in F2 and F3, which matched the ORAC values of F1. In the case of the purple Hindukusch genotype, the highest AC was detected in F2. This fraction contained the highest amount of free PCs, including anthocyanins, whose AC exceeded that of the components in F1 and F3. Finally, F1 of Annapurna (equivalent to F2 of the genotypes containing husks) showed the highest AC, and this value decreased going toward the inner part of the grain as did free, bound and total PCs. Based on these results, the 0–15% outer fractions of the genotypes containing husks and the 0–5% fraction of Annapurna made significant contributions to the total antioxidant capacity of all whole flours.

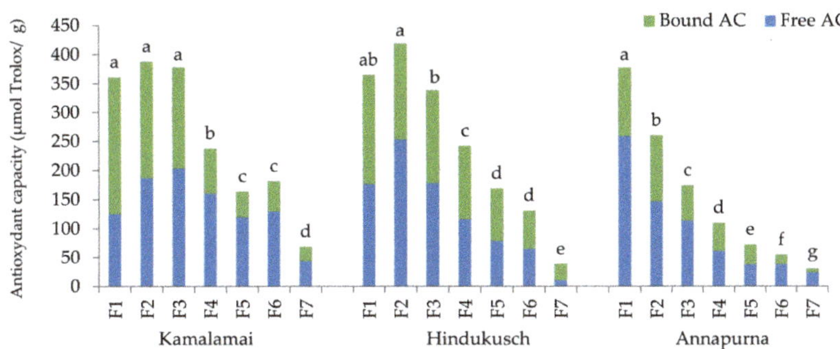

Figure 5. Antioxidant capacity (μmol Trolox/g) detected in seven pearling fractions in three food barley genotypes. Results are presented as the mean; different letters indicate significant differences between pearled fractions of each genotype on log-transformed data; (Tukey-Kramer's HSD for α = 0.05).

3.6. Association between Variables

Figure 6 shows the PCA biplot of ash, protein, bioactive compounds and antioxidant capacity in the seven pearling fractions of the three food barley genotypes. The first two principal component axes explained more than two thirds of the total variability in the standardized data from Figures 1–3 and and Tables 4 and 5. The first axis, explaining 50% of the total variation, seemed to be related to the contrast between the bioactive compounds found in the outer fractions (ash, arabinoxylans, phenolic compounds and tocotrienols) and the β-glucans present in the endosperm. The second, explaining 17% of the total, seemed to be related to compounds such as tocopherols, protein and minor phenolic compounds like catechins (Cat) and bound hydroxybenzoic acid (OHB) found in intermediate layers.

The size of the circles for each pearling fractions was proportional to their total antioxidant capacity and clearly descends from fraction F1 to F7 for all genotypes. The distribution pattern was similar for all genotypes, with little differences for the outermost fractions. The size of the blue squares representing the 28 phenolic compounds was proportional to their content within the fractions. Overall, Hindukusch, with the outermost fractions being further away from the origin of coordinates, seemed higher in such bioactive compounds as tocotreinols, prodelphinidin B4 (PdB4) bound ferulic acid (B-F), and anthocyanins (AN) than the other two genotypes. Highlighted in quadrant II, bound *p*-coumaric acid (*p*-Cm) showed higher content in the fractions that mainly contain husks, and in quadrant I the bound ferulic (B-F) and diferulic (B-DiF) acids, and the procyanidin B3 (PcB3) and prodelphinidin B4 (PdB4) present in the outer external fraction.

Whereas PCA is an extremely powerful tool for the visualization of multidimensional data, it does not allow for statistical inferences about contents across grain sections. These comparisons are shown in Figure 7, which summarizes the Tukey's HSD Mean Comparison groupings for four sequential spatial sections of the barley grain, determined by differential aggregation of the seven pearling fraction from each of the three, hulled, partially-hull-less and hull-less food barley genotypes used. The husks had a high content in ash, arabinoxylans, some specific major phenolic compounds such as bound *p*-coumaric, ferulic and diferulic acids, procyanidin B3 and prodelphinidin B4, resulting in a high antioxidant capacity, similar to that of the outermost layers. Current regulations require removal of the inedible husks from hulled barley to be used for human food [46]. However, due to its high content of bioactive compounds, either the whole hulled barley grains or the external husks of food hulled barley genotypes should be considered as a valuable material for animal feeds. Apart from providing the same concentration of bioactive compounds as husks, the outermost layers also had high contents of tocols. Therefore, the outermost layers of naked barley or previously de-husked barley for human food represent the most valuable source of bioactive compounds. The intermediate layers, provide high contents of protein, β-glucans, tocopherols and some phenolic compounds such as catechins and hydroxybenzoic acid. Finally, the endosperm has the highest β-glucans contents and relative high presence of catechins and hydroxybenzoic acid.

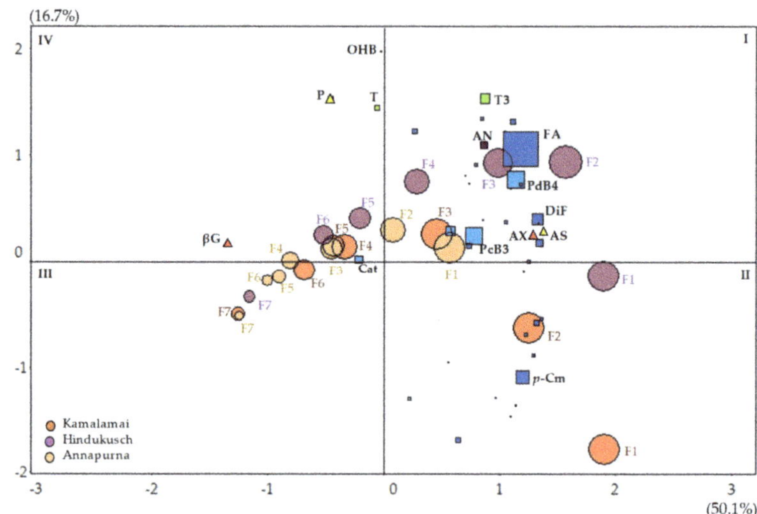

Figure 6. Biplot of the Principal Component Analysis of ash, protein, dietary fiber and 60 bioactive compounds in seven sequential pearling fractions in three food barley genotypes. The size of the squares is proportional to each phenolic compound content; Circle size is proportional to total antioxidant capacity. F: fraction; AS: Ash; P: Protein; βG: β-glucans; AX: arabinoxylans; T: tocopherols; T3: tocotrienols. The following eight main phenolic compounds are also labeled: FA: bound ferulic acid; p-Cm: bound p-coumaric acid; DiF: bound diferulic acid; Cat: catechins; OHB: m- or o-hydroxybenzoic acid; PcB3: procyanindin B3, PdB4: prodelfinidin B4; Light blue squares: flavone glycosides; AN: total anthocyanins.

	K	H	A	AS	P	βG	AX	T‡	T3‡	Cat‡	PcB3‡	PdB4‡	OHB‡	pCm‡	FA‡	DiF‡	AC‡
Husks	F1 F2	 F1		a	c	c	a	b	b	b	a	a	b	a	a	a	a
Outermost Layers	F3 F4	F2 F3	F1 F2	b	b	b	a	a	a	a	a	a	a	b	a	a	a
Intermediate Layers	F5 F6	F4 F5	F3 F4	c	a	a	b	a	b	ab	b	b	a	c	b	b	b
Endosperm	 F7	F6 F7	F5 F6 F7	d	c	a	c	c	c	b	c	c	b	d	c	c	c

Figure 7. Tukey's HSD groupings for a number of compounds for the different sections of the grains across three food barley genotypes. The colored left columns represent the different pearling fractions (F1–F7) of the Kamalamai (K), Hindukusch (H) and Annapurna (A) used for the aggregation into four grain sections shown in the first column. A: Ash; P: protein; βG: β-glucans; AX: arabinoxylans; T: tocopherols; T3: tocotrienols; Cat: catechins; PB3: Procyanidin B3; PB4: prodelphinidin B4; OHB: m- or o-hydroxybenzoic acid; pCm: bound p-coumaric acid, FA: bound ferulic acid; DiF: bound diferulic acid; AC: total antioxidant capacity. ‡ Analyses carried out on log-transformed data.

In conclusion, high β-glucan food barley genotypes can be an excellent source of not just of dietary fiber but a plethora of phenolic compounds with potential health promoting properties. Whole or lightly pearled grains, as well as their specific pearling fractions, could be used as a diverse source of valuable functional ingredients.

Supplementary Materials: The following are available online at https://www.mdpi.com/2304-8158/10/3/565/s1, Table S1: Tocopherol and Tocotrienol contents in pearling fractions of three barley genotypes; Table S2: Anthocyanins contents (µg/g) in the pearling fractions of the partially-hull-less and purple genotype.

Author Contributions: Conceptualization, M.M. and M.-P.R.; methodology, M.M.-S., M.M. and M.-P.R.; formal chemical analyses, M.M.-S., A.M. and E.P.; statistical analyses and data curation, M.M.-S. and I.R.; writing—original draft preparation, M.M.-S.; writing—review and editing, I.R., M.-P.R. and M.M.; supervision, M.M.; funding acquisition, I.R. and M.M. All authors have read and agreed to the published version of the manuscript.

Funding: This work was funded by project AGL 2015-69435-C3-1 from the Spanish Ministry of Economy and Competitiveness. Mariona Martínez-Subirà was supported by a pre-doctoral fellowship (BES-2016-078654/AGL 2015-69435-C3-1).

Institutional Review Board Statement: Not applicable.

Informed Consent Statement: Not applicable.

Data Availability Statement: Data will be available at https://dataverse.csuc.cat/dataverse/Agrotecnio, accessed on 30 October 2020.

Acknowledgments: Our thanks to Semillas Batlle, SA (Bell-lloc d'Urgell, Lleida, Spain) for providing the food barley genotypes used in this work.

Conflicts of Interest: The authors declare no conflict of interest.

References

1. Faostat. Available online: http://faostat.fao.org (accessed on 20 January 2021).
2. Baik, B.-K.; Ullrich, S.E. Barley for food: Characteristics, improvement, and renewed interest. *J. Cereal Sci.* **2008**, *48*, 233–242. [CrossRef]
3. FDA. Food Labeling: Health Claims, Soluble Dietary Fiber from Certain Foods and Coronary Heart Disease. Available online: https://www.federalregister.gov/documents/2008/02/25/E8-3418/food-labeling-health-claims-soluble-fiber-from-certain-foods-and-risk-of-coronary-heart-disease (accessed on 30 October 2020).
4. EFSA. Scientific Opinion on the Substantiation of Health Claims Related to β-Glucans from Oats and Barley and Maintenance of Normal Blood LDL-Cholesterol Concentrations (ID 1236, 1299), Increase in Satiety Leading to a Reduction in Energy Intake (ID 851, 852), Reduction of Post-Prandial Glycaemic Responses (ID 821, 824), and "Digestive Function" (ID 850) Pursuant to Article 13 (1) of Regulation (EC) No 1924/2006. *EFSA J.* **2011**, *9*, 2207.
5. De Graaff, P.; Govers, C.; Wichers, H.J.; Debets, R. Consumption of β-glucans to spice up T cell treatment of tumors: A review. *Expert Opin. Biol. Ther.* **2018**, *18*, 1023–1040. [CrossRef] [PubMed]
6. Hong, F.; Yan, J.; Baran, J.T.; Allendorf, D.J.; Hansen, R.D.; Ostroff, G.R.; Xing, P.X.; Cheung, N.-K.V.; Ross, G.D. Mechanism by Which Orally Administered β-1,3-Glucans Enhance the Tumoricidal Activity of Antitumor Monoclonal Antibodies in Murine Tumor Models. *J. Immunol.* **2004**, *173*, 797–806. [CrossRef]
7. Bai, J.; Ren, Y.; Li, Y.; Fan, M.; Qian, H.; Wang, L.; Wu, G.; Zhang, H.; Qi, X.; Xu, M.; et al. Physiological functionalities and mechanisms of β-glucans. *Trends Food Sci. Technol.* **2019**, *88*, 57–66. [CrossRef]
8. Izydorczyk, M.; Dexter, J. Barley β-glucans and arabinoxylans: Molecular structure, physicochemical properties, and uses in food products—A Review. *Food Res. Int.* **2008**, *41*, 850–868. [CrossRef]
9. Fadel, A.; Mahmoud, A.M.; Ashworth, J.J.; Li, W.; Ng, Y.L.; Plunkett, A. Health-related effects and improving extractability of cereal arabinoxylans. *Int. J. Biol. Macromol.* **2018**, *109*, 819–831. [CrossRef]
10. Malunga, L.N.; Beta, T. Antioxidant Capacity of Water-Extractable Arabinoxylan from Commercial Barley, Wheat, and Wheat Fractions. *Cereal Chem. J.* **2015**, *92*, 29–36. [CrossRef]
11. Gangopadhyay, N.; Rai, D.K.; Brunton, N.P.; Gallagher, E.; Hossain, M.B. Antioxidant-guided isolation and mass spectrometric identification of the major polyphenols in barley (*Hordeum vulgare*) grain. *Food Chem.* **2016**, *210*, 212–220. [CrossRef] [PubMed]
12. Martínez, M.; Motilva, M.J.; López de las Hazas, M.C.; Romero, M.P.; Vaculova, K.; Ludwig, I.A. Phytochemical composition and β-glucan content of barley genotypes from two different geographic origins for human health food production. *Food Chem.* **2018**, *245*, 61–70. [CrossRef]

13. Abdel-Aal, E.-S.M.; Choo, T.-M.; Dhillon, S.; Rabalski, I. Free and Bound Phenolic Acids and Total Phenolics in Black, Blue, and Yellow Barley and Their Contribution to Free Radical Scavenging Capacity. *Cereal Chem. J.* **2012**, *89*, 198–204. [CrossRef]
14. Suriano, S.; Iannucci, A.; Codianni, P.; Fares, C.; Russo, M.; Pecchioni, N.; Marciello, U.; Savino, M. Phenolic acids profile, nutritional and phytochemical compounds, antioxidant properties in colored barley grown in southern Italy. *Food Res. Int.* **2018**, *113*, 221–233. [CrossRef]
15. Corrêa, T.A.F.; Rogero, M.M.; Hassimotto, N.M.A.; Lajolo, F.M. The Two-Way Polyphenols-Microbiota Interactions and Their Effects on Obesity and Related Metabolic Diseases. *Front. Nutr.* **2019**, *6*, 188. [CrossRef]
16. Panfili, G.; Fratianni, A.; Di Criscio, T.; Marconi, E. Tocol and β-glucan levels in barley varieties and in pearling by-products. *Food Chem.* **2008**, *107*, 84–91. [CrossRef]
17. Badea, A.; Carter, A.; Legge, W.G.; Swallow, K.; Johnston, S.P.; Izydorczyk, M.S. Tocols and oil content in whole grain, brewer's spent grain, and pearling fractions of malting, feed, and food barley genotypes. *Cereal Chem. J.* **2018**, *95*, 779–789. [CrossRef]
18. Irakli, M.; Lazaridou, A.; Mylonas, I.; Biliaderis, C.G. Bioactive Components and Antioxidant Activity Distribution in Pearling Fractions of Different Greek Barley Cultivars. *Foods* **2020**, *9*, 783. [CrossRef]
19. Sharma, P.; Gujral, H.S. Cookie making behavior of wheat-barley flour blends and effects on antioxidant properties. *LWT* **2014**, *55*, 301–307. [CrossRef]
20. Martínez-Subirà, M.; Romero, M.P.; Puig, E.; Macià, A.; Romagosa, I.; Moralejo, M. Purple, high β-glucan, hulless barley as valuable ingredient for functional food. *LWT* **2020**, *131*, 109582. [CrossRef]
21. Marconi, E.; Graziano, M.; Cubadda, R. Composition and Utilization of Barley Pearling By-Products for Making Functional Pastas Rich in Dietary Fiber and β-Glucans. *Cereal Chem. J.* **2000**, *77*, 133–139. [CrossRef]
22. Blandino, M.; Locatelli, M.; Sovrani, V.; Coïsson, J.D.; Rolle, L.; Travaglia, F.; Giacosa, S.; Bordiga, M.; Scarpino, V.; Reyneri, A.; et al. Progressive Pearling of Barley Kernel: Chemical Characterization of Pearling Fractions and Effect of Their Inclusion on the Nutritional and Technological Properties of Wheat Bread. *J. Agric. Food Chem.* **2015**, *63*, 5875–5884. [CrossRef] [PubMed]
23. Madhujith, T.; Shahidi, F. Antioxidant potential of barley as affected by alkaline hydrolysis and release of insoluble-bound phenolics. *Food Chem.* **2009**, *117*, 615–620. [CrossRef]
24. Thiex, N.; Novotny, L.; Crawford, A. Determination of Ash in Animal Feed: AOAC Official Method 942.05 Revisited. *J. AOAC Int.* **2012**, *95*, 1392–1397. [CrossRef] [PubMed]
25. Serra, A.; Rubio, L.; Macià, A.; Valls, R.-M.; Catalán, Ú.; De La Torre, R.; Motilva, M.-J. Application of dried spot cards as a rapid sample treatment method for determining hydroxytyrosol metabolites in human urine samples. Comparison with microelution solid-phase extraction. *Anal. Bioanal. Chem.* **2013**, *405*, 9179–9192. [CrossRef] [PubMed]
26. Huang, D.; Ou, B.; Hampsch-Woodill, M.; Flanagan, J.A.; Prior, R.L. High-Throughput Assay of Oxygen Radical Absorbance Capacity (ORAC) Using a Multichannel Liquid Handling System Coupled with a Microplate Fluorescence Reader in 96-Well Format. *J. Agric. Food Chem.* **2002**, *50*, 4437–4444. [CrossRef]
27. Swanston, J.S.; Ellis, R.P.; Pérez-Vendrell, A.; Voltas, J.; Molina-Cano, J.-L. Patterns of Barley Grain Development in Spain and Scotland and Their Implications for Malting Quality. *Cereal Chem. J.* **1997**, *74*, 456–461. [CrossRef]
28. Andersson, A.A.M.; Lampi, A.-M.; NyströmL; Piironen, V.; Li, L.; Ward, J.L.; Gebruers, K.; Courtin, C.M.; Delcour, J.A.; Boros, D.; et al. Phytochemical and Dietary Fiber Components in Barley Varieties in the Healthgrain Diversity Screen. *J. Agric. Food Chem.* **2008**, *56*, 9767–9776. [CrossRef]
29. Izydorczyk, M.S.; McMillan, T.; Bazin, S.; Kletke, J.; Dushnicky, L.; Dexter, J.; Chepurna, A.; Rossnagel, B. Milling of Canadian oats and barley for functional food ingredients: Oat bran and barley fibre-rich fractions. *Can. J. Plant Sci.* **2014**, *94*, 573–586. [CrossRef]
30. Izydorczyk, S.M.; Biliaderis, C.G. Arabinoxylans: Technologically and Nutritionally Functional Plant Polysaccharides. In *Functional Food Carbohydrates*; Biliaderis, C.G., Izydorczyk, S.M., Eds.; Taylor & Francis Group: Boca Raton, FL, USA, 2007; pp. 250–290.
31. Hassan, A.S.; Houston, K.; Lahnstein, J.; Shirley, N.; Schwerdt, J.G.; Gidley, M.J.; Waugh, R.; Little, A.; Burton, R.A. A Genome Wide Association Study of arabinoxylan content in 2-row spring barley grain. *PLoS ONE* **2017**, *12*, e0182537. [CrossRef] [PubMed]
32. Cavallero, A.; Gianinetti, A.; Finocchiaro, F.; Delogu, G.; Stanca, A.M. Tocols in hull-less and hulled barley genotypes grown in contrasting environments. *J. Cereal Sci.* **2004**, *39*, 175–180. [CrossRef]
33. Lampi, A.-M.; Nurmi, T.; Ollilainen, V.; Piironen, V. Tocopherols and Tocotrienols in Wheat Genotypes in the Healthgrain Diversity Screen. *J. Agric. Food Chem.* **2008**, *56*, 9716–9721. [CrossRef]
34. Moreau, R.A.; Flores, R.A.; Hicks, K.B. Composition of Functional Lipids in Hulled and Hulless Barley in Fractions Obtained by Scarification and in Barley Oil. *Cereal Chem. J.* **2007**, *84*, 1–5. [CrossRef]
35. Idehen, E.; Tang, Y.; Sang, S. Bioactive phytochemicals in barley. *J. Food Drug Anal.* **2017**, *25*, 148–161. [CrossRef]
36. Graebner, R.C.; Wise, M.; Cuesta-Marcos, A.; Geniza, M.; Blake, T.; Blake, V.C.; Butler, J.; Chao, S.; Hole, D.J.; Horsley, R.; et al. Quantitative Trait Loci Associated with the Tocochromanol (Vitamin E) Pathway in Barley. *PLoS ONE* **2015**, *10*, e0133767. [CrossRef]
37. Gamel, T.; Abdel-Aal, E.-S.M. Phenolic acids and antioxidant properties of barley wholegrain and pearling fractions. *Agric. Food Sci.* **2012**, *21*, 118–131. [CrossRef]
38. Holtekjølen, A.K.; Sahlstrøm, S.; Knutsen, S.H. Phenolic contents and antioxidant activities in covered whole-grain flours of Norwegian barley varieties and in fractions obtained after pearling. *Acta Agric. Scand. Sect. B Plant Soil Sci.* **2011**, *61*, 67–74. [CrossRef]

39. Giordano, D.; Reyneri, A.; Locatelli, M.; Coïsson, J.D.; Blandino, M. Distribution of bioactive compounds in pearled fractions of tritordeum. *Food Chem.* **2019**, *301*, 125228. [CrossRef] [PubMed]
40. Gangopadhyay, N.; Harrison, S.M.; Brunton, N.P.; Hidalgo-Ruiz, J.L.; Gallagher, E.; Rai, D.K. Brans of the roller-milled barley fractions rich in polyphenols and health-promoting lipophilic molecules. *J. Cereal Sci.* **2018**, *83*, 213–221. [CrossRef]
41. Lee, C.; Han, D.; Kim, B.; Baek, N.; Baik, B.-K. Antioxidant and anti-hypertensive activity of anthocyanin-rich extracts from hulless pigmented barley cultivars. *Int. J. Food Sci. Technol.* **2013**, *48*, 984–991. [CrossRef]
42. Zhang, X.-W.; Jiang, Q.-T.; Wei, Y.-M.; Liu, C. Inheritance analysis and mapping of quantitative trait loci (QTL) controlling individual anthocyanin compounds in purple barley (*Hordeum vulgare* L.) grains. *PLoS ONE* **2017**, *12*, e0183704. [CrossRef] [PubMed]
43. Yoshida, A.; Sonoda, K.; Nogata, Y.; Nagamine, T.; Sato, M.; Oki, T.; Hashimoto, S.; Ohta, H. Determination of Free and Bound Phenolic Acids, and Evaluation of Antioxidant Activities and Total Polyphenolic Contents in Selected Pearled Barley. *Food Sci. Technol. Res.* **2010**, *16*, 215–224. [CrossRef]
44. Zhu, Y.; Li, T.; Fu, X.; Abbasi, A.M.; Zheng, B.; Liu, R.H. Phenolics content, antioxidant and antiproliferative activities of dehulled highland barley (*Hordeum vulgare* L.). *J. Funct. Foods* **2015**, *19*, 439–450. [CrossRef]
45. Xia, X.; Xing, Y.; Kan, J. Antioxidant activity of Qingke (highland hull-less barley) after extraction/hydrolysis and in vitro simulated digestion. *J. Food Process. Preserv.* **2019**, *44*, 14331. [CrossRef]
46. FDA. Draft Guidance for Industry and FDA Staff: Whole Grain Label Statements (Docket No FDA-2006-D-0298). Issued by: Center for Food Safety and Applied Nutrition. Available online: https://www.regulations.gov/docket?D=FDA-2006-D-0298 (accessed on 21 January 2021).

 *foods*

Article

Concentration of Potentially Bioactive Compounds in Italian Extra Virgin Olive Oils from Various Sources by Using LC-MS and Multivariate Data Analysis

Anna Różańska [1,2], Marina Russo [1,*], Francesco Cacciola [3], Fabio Salafia [1], Żaneta Polkowska [2], Paola Dugo [1,4] and Luigi Mondello [1,4,5,6]

1   Department of Chemical, Biological, Pharmaceutical and Environmental Sciences, University of Messina, 98122 Messina, Italy; anna.rozanska@pg.edu.pl (A.R.); fsalafia@unime.it (F.S.); pdugo@unime.it (P.D.); lmondello@unime.it (L.M.)
2   Department of Analytical Chemistry, Faculty of Chemistry, Gdańsk University of Technology, 80-233 Gdańsk, Poland; zanpolko@pg.edu.pl
3   Department of Biomedical, Dental, Morphological and Functional Imaging Sciences, University of Messina, 98122 Messina, Italy; cacciolaf@unime.it
4   Chromaleont s.r.l., c/o Department of Chemical, Biological, Pharmaceutical and Environmental Sciences, University of Messina, 98122 Messina, Italy
5   Department of Sciences and Technologies for Human and Environment, University Campus Bio-Medico of Rome, 00128 Rome, Italy
6   BeSep s.r.l., c/o Department of Chemical, Biological, Pharmaceutical and Environmental Sciences, University of Messina, 98122 Messina, Italy
*   Correspondence: marina.russo@unime.it; Tel.: +39-090-676-6567

Received: 25 June 2020; Accepted: 10 August 2020; Published: 13 August 2020

Abstract: High quality extra virgin olive oils represent an optimal source of nutraceuticals. The European Union (EU) is the world's leading olive oil producer, with the Mediterranean region as the main contributor. This makes the EU the greatest exporter and consumer of olive oil in the world. However, small olive oil producers also contribute to olive oil production. Beneficial effects on human health of extra virgin olive oil are well known, and these can be correlated to the presence of vitamin E and phenols. Together with the origin of the olives, extraction technology can influence the chemical composition of extra virgin olive oil. The aim of this study was to investigate the concentration of potentially bioactive compounds in Italian extra virgin olive oils from various sources. For this purpose, vitamin E and phenolic fractions were characterized using high-performance liquid chromatography (HPLC) coupled with fluorescence, photodiode array and mass spectrometry detection in fifty samples of oil pressed at industrial plants and sixty-six samples of oil produced in low-scale mills. Multivariate statistical data analysis was used to determine the applicability of selected phenolic compounds as potential quality indicators of extra virgin olive oils.

Keywords: extra-virgin olive oils; oleuropein; phenols; tocopherols; HPLC-MS; multivariate statistical data

---

1. Introduction

As reported by the European Commission [1] more than 69% of the world's olive oil comes from the Mediterranean region, primarily from Spain. This makes the European Union (EU) the leading producer, exporter and consumer of olive oil in the world. The EU's major producer member states are: Spain, Italy, Greece, France, Portugal, Croatia, Slovenia, Malta and Cyprus. However, between these nine nations Spain is the biggest producer, with 63% of the EU's total production. In addition, a substantial contribution to the total EU olive oil production comes from Italy, Greece and Portugal, with a production of 17%, 14% and 5%, respectively. Extra virgin olive oil (EVOO) produced in the EU is subjected to marketing standards. These requirements ensure that consumers receive a product with a standardized and satisfactory quality, and on the other hand ensure equal economic conditions to EU producers [2].

Parallel to the EU industrial EVOO production there is a market of smaller olive oil producers. This kind of manufacturing respects all the requirements demanded by the EU. However, the smaller olive oil producers try to distinguish their extra virgin olive oils from the anonymity of industrial oil production.

Beneficial effects on human health of EVOO are well known, and these can be correlated with the presence of a well-balanced acidic composition, and especially to the presence of fat-soluble vitamin and hydrophilic phenolic compounds (vitamin E and phenols) [3–6]. Vitamin E is composed of eight isomers, and in EVOO is principally represented by α-tocopherol [7], that is the isomer with the greatest biological activity [8,9]. Hydrophilic phenolic compounds present in EVOO belong to different classes: secoiridoids, phenolic acids, phenolic alcohols, lignans and flavonoids [10]. Thanks to this qualitative bioactive molecule composition, EVOO can be considered as a functional food. Moreover, together with the health promoting qualities these molecules prolong the shelf life of the oil increasing its stability, and stabilize the oil during frying [11,12].

In literature, several research articles can be found concerning the characterization of bioactive molecules in EVOOs [13–15], and some of these reported their determination despite being applied to limited data samples [7,16–23]. Recently our research group published a research article on the content of bioactive antioxidant molecules among 186 high quality Italian extra virgin olive oils belonging to eleven regions [24]. Data obtained showed that there were some differences between different regions even though they were not so marked and significant; this highlighted how the cold extraction technology employed kept antioxidant content unaltered.

In other cases, olive extraction technology was found as a parameter which could influence the chemical composition of EVOOs [25]. Anyway, all the stages of the oil production can effect the bioactive molecule composition. For instance, parameters that influence the quality and quantity of phenols and vitamin E are olive collection, type of crusher, crushing speed and conditions, type of decanter and its regulation, and so on [26]. So, differences in antioxidant molecule composition not only depend on cultivar, quality, territory, climate, storage conditions, olive maturity index, but also on production techniques [16,26].

The aim of this study was to investigate the nutraceutical properties of Italian extra virgin olive oils from various sources, namely, small farms and local distribution centers. For this purpose, vitamin E and phenolic fractions were characterized using HPLC techniques coupled with fluorescence (FLD), photodiode array (PDA) and mass spectrometry (MS) detection in fifty samples of oil pressed at industrial plants and sixty-six samples of oil produced in low-scale mills. Multivariate statistical data analysis was used to determine the applicability of selected phenolic compounds as potential quality indicators of extra virgin olive oils.

To the best of our knowledge, this is the first research article that reports the bioactive molecule content and a principal components analysis (PCA) statistical correlation between vitamin E and phenols in a large array of Italian high quality EVOOs (66) produced in different regions and fifty commercially available EVOOs.

2. Materials and Methods

2.1. Materials and Reagents

Standards: α-, β-, γ-tocopherol and α-tocotrienol were purchased from Extrasynthese (Extrasynthese S.A., Genay Cedex, France); gallic acid, tyrosol, hydroxytyrosol, apigenin, luteolin, oleuropein and verbascoside were purchased from Merck Life Science (Merck KGaA, Darmstadt, Germany) as was the internal standard (ethyl gallate, IS). Solvents: acetonitrile, ethanol, formic acid, isopropanol, methanol and *n*-hexane used for chromatographic analysis were purchased from Merck Life Science (Merck KGaA, Darmstadt, Germany). Water (resistivity above 18 MΩcm) was obtained from a Milli-Q SP Reagent Water System (Merck KGaA, Darmstadt, Germany).

2.2. Samples

Sixty six Italian extra-virgin olive oils (marked as EVOOs) were collected from various Italian low-scale oil mills sited in: Abruzzo, Apulia, Calabria, Campania, Lazio, Liguria, Garda area (Lombardy, Trentino, Veneto), Sardinia, Sicily, Tuscany and Umbria. These samples were collected on the basis of the European extra quality olive oil award called "il Magnifico". This award is assigned to the best producers of extra quality olive oil in Europe. Moreover, fifty commercially available extra-virgin olive oils (marked as COOs) were purchased from local shops and supermarkets labelled as samples from Italy, EU countries, EU/Italy and Sicily. As reported in commercial olive oil labels, both olive oil samples from EU countries and EU/Italy were processed by olive mills located in Italy, but olives were harvested from EU member states or Italy and other EU member states respectively. All information about the samples are listed in Table 1. The oils were poured into amber glass vials and stored at −20 °C, each sample was thawed and analyzed on the same day.

Table 1. High quality and commercial [§] extra-virgin olive oils analyzed.

Country	Denomination	Label	*Cultivar*, Year (n. Samples)
Apulia	PDO	Terra di Bari	*Coratina*, 2019 (1)
	MV		*Peranzana*, 2019 (2)
			Olivastra, 2019 (1)
			Coratina, 2019 (2)
			Picholine, 2019 (1)
	Blend		*Coratina-Peranzana*, 2019 (1)
Sicily	PDO	Monti Iblei	*Tonda Iblea-Moresca*, 2018 (1)
		Valle del Belice	*Nocellara del Belice*, 2018 (1) [§]
		Val di Mazara	*Biancolilla-Cerasuola-Nocellara del Belice*, 2018 (2) [§]
		Valli Trapanesi	*Biancolilla-Cerasuola-Nocellara del Belice*, 2018 (1) [§]
	PGI		*Cerasuola*, 2018 (1) [§]
			Nocellara-Biancolilla-Cerasuola, 2018 (3) [§]
			Nocellara Etnea, 2018 (1) [§]
			Nocellara del Belice, 2018 (1) [§]
	MV		*Nocellara del Belice*, 2019 (1)
			Nocellara Etnea, 2018 (1) [§]
Tuscany	PDO	Chianti classico	*Moraiolo*, 2019 (2)
			Leccino-Moraiolo-Frantoio, 2019 (3)
			Moraiolo-Frantoio, 2019 (2)
	PGI		*Frantoio*, 2019 (1)
			Leccino-Moraiolo-Frantoio, 2019 (5)
			Leccino-Moraiolo-Frantoio- Pendolino, 2019 (1)
	Bio		*Leccino-Moraiolo-Frantoio*, 2019 (5)
			Moraiolo, 2019 (2)
			Frantoio, 2019 (2)
	Blend		*Leccino-Moraiolo-Frantoio-Pendolino*, 2019 (2)
	MV		*Frantoio*, 2019 (1)
Liguria	PDO	Riviera Ligure	*Taggiasca*, 2019 (1)
	MV		*Taggiasca*, 2019 (1)

Table 1. *Cont.*

Country	Denomination	Label	*Cultivar*, **Year (n. Samples)**
Campania	MV		*Leccio del Corno*, 2019 (2)
Abruzzo	PDO	Colline Teatine	*Gentile di Chieti-Intosso-Leccino*, 2018 (1)
			Dritta, 2019 (1)
			Intosso, 2019 (2)
	Blend		*Gentile di Chieti-Intosso-Leccino*, 2019 (2)
Garda area	PDO	Garda Trentino	*Casaliva-Leccino-Frantoio*, 2019 (1)
	MV		*Casaliva*, 2019 (2)
	Blend		*Casaliva-Frantoio-Leccino*, 2019 (1)
Calabria	Bio		*Carolea*, 2019 (1)
	MV		*Ottobratica*, 2019 (1)
	Blend		*Ottobratica-Sinopolese*, 2019 (1)
Lazio	Bio		*Canino*, 2019 (1)
			Fratoio, 2018 (1)
	MV		*Itrana*, 2018 (1)
			Leccino, 2019 (1)
			Rosciola, 2018 (1)
	Blend		*Caninese-Frantoio-Maurino-Leccino-Pendolino*, 2019 (1)
Sardinia	PDO	Sardegna	*Bosana-Semidana*, 2019 (1)
			Bosana, 2019 (1)
	Blend		*Bosana-Frantoio-Semidana-Coratina-Leccino*, 2019 (2)
	MV		*Bosana*, 2019 (2)
Umbria	Blend		*Leccino-Frantoio-Moraiolo*, 2019 (1)
			Leccino-Frantoio-Moraiolo-S.Felice,2019 (1)
Italy	Blend	EVOO	Not reported, 2018 (15) [§]
EU	Olives from EU member states	EVOO milled in Italy	Not reported, 2018 (19) [§]
Italy + EU	Olives from Italy + EU member states	EVOO milled in Italy	Not reported, 2018 (4) [§]

[§]: commercial extra virgin olive oil samples; PDO: protected designation of origin; PGI: protected geographical identification; Bio: organic farming; MV: monovarietal.

2.3. Tocopherols and Tocotrienols Determination by NP-HPLC-FLD

Tocopherols and tocotrienols were determined according to the method previously developed and validated by Dugo et al. [24]. Briefly, olive oils samples were diluted in n-hexane (1:10, 1:15 or 1:30 *v/v*). The HPLC analyses were carried out using a Shimadzu Nexera-X2 system (Shimadzu, Milan, Italy) equipped with an on-line degasser (DGU-20ASR), an autosampler (SIL-30 AC), two dual-plunger parallel-flow pumps (LC-30 AD), a column oven (CTO-20AC), a communication bus module (CBM-20A) and a fluorescence detector (RF-20AXS). Phenolic components were separated on a normal-phase Ascentis Si column (250 × 4.6 mm I.D. with 5 μm particle size, Merck KGaA, Darmstadt, Germany) under the following conditions: mobile phase n-hexane:isopropanol (99:1; *v/v*); flow rate (1.7 mg L^{-1}); column oven temperature (25 °C) and injection volume (5 μL). The identification and quantification were conducted by using a fluorescence detector at an excitation wavelength of 290 nm and emission wavelength of 330 nm. Data acquisition was performed by the LCMS solution v. 5.85 software (Shimadzu, Milan, Italy).

To quantify the vitamin E content in the EVOOs sample calibration curves have been constructed by using each single available standard, according to the method previously developed and validated by Dugo et al. [24]. Briefly, five different concentrations of each component, in the range between 5 and 0.005 mg L^{-1}, prepared by diluting a stock solution of about 100 mg L^{-1}, were analyzed five consecutive times by NP-HPLC.

2.4. Phenols Determination by RP-HPLC-PDA/MS

Phenolic acid, phenolic alcohols, flavonoids and secoiridoids were determined using the method described by Dugo et al. [24]. Briefly, 1 mL of olive oil was diluted in 1 mL of n-hexane and homogenized. Next, the sample was extracted with 1 mL of H_2O:methanol (2:3, v/v) solution for 2 min in an Elmasonic P 60H ultrasonic bath (Elma Schmidbaure GmbH, Singen, Germany), and centrifuged for 10 min at 3000 rpm. The polar phase was transferred and washed with 1 mL of n-hexane. To the final extract, 20 µL of ethyl gallate (1 mg mL^{-1} in methanol) was added. Sample extraction methods validation was carried out according to Dugo et al. [24] by using a sample of soybean oil added of known amounts of luteolin, oleuropein, apigenin, tyrosol and hydroxytyrosol.

The HPLC analyses were carried out using a Shimadzu Prominence LC-20A (Shimadzu, Kyoto, Japan) equipped with a degasser (DGU-20A5), an autosampler (SIL-20 A), two dual-plunger parallel-flow pumps (LC-20 AD), a communication bus module (CBM-20A), a photodiode array detector (SPD-M20A) and a single-quadrupole mass spectrometer (LCMS-2020) equipped with an electrospray ionization (ESI) source, operating in negative ionization mode. The separation was conducted on a reverse-phase Ascentis Express C18 column (150 × 4.6 mm, 2.7 µm, Merck Life Science, Merck KGaA, Darmstadt, Germany) under the following conditions: mobile phases A—0.1% acetic acid and B—acetonitrile with 0.1% formic acid; flow rate (1 mL min^{-1}); column oven temperature (25 °C) and injection volume (5 µL). The gradients used were as follows: 0 min, 10% B; 4 min, 35% B; 12 min, 47% B; 12.5 min, 60% B; 16 min, 75% B; and 21 min, 100% B. Identification was carried out by both PDA (280 nm) and MS detection under the following conditions: mass spectral range (m/z 100–800); event time (0.5 s); nebulizing gas flow, N_2 (1.5 L min^{-1}); drying gas flow, N_2 (5 L min^{-1}); heat block temperature (300 °C); and desolvation line (DL) temperature (280 °C). Single ion monitoring (SIM) was used for quantification: gallic acid (170 m/z), tyrosol (138 m/z), hydroxytyrosol (154 m/z), apigenin (270 m/z), luteolin (286 m/z), oleocanthal (304 m/z), oleacein (320 m/z), oleuropein aglycone (378 m/z) and ligstroside aglycone (362 m/z). Data acquisition was performed by the LCMS solution v. 5.85 software (Shimadzu, Kyoto, Japan).

To quantify the hydrophilic phenols content in the one hundred and eighty-six high quality extra-virgin olive oil samples, the method previously described by Dugo et al. [24] was used. Briefly, five different concentrations of each component, in the range between 100 and 0.1 mg L^{-1}, prepared by diluting a stock solution of about 1000 mg L^{-1} were analyzed five consecutive times by RP-HPLC. Before injection, 20 µL of internal standard (IS) ethyl gallate (1 mg mL^{-1}) was added to 1 mL of each standard solution.

2.5. Multivariate Statistical Analysis

Each sample was analyzed in triplicate. One-way ANOVA using Tukey's test was applied to evaluate the significant differences at a level of $p < 0.05$ among means (Statistica 12, StatSoft, Inc., Tulsa, OK, USA). The determined phenol concentration values in olive oil samples were used as input data for multivariate statistical data analysis using the dedicated Python toolset Orange v. 3.13 (Bioinformatics Lab, University of Ljubljana, Ljubljana, Slovenia). Initial data processing included standardization (centering by a mean value and scaling by standard deviation). Missing data (determined concentration below LOQ) was replaced by respective LOD/3 values. Analysis of variance within variables and feature selection was performed using ANOVA. During the research, various chemometric approaches were used, namely hierarchical cluster analysis (HCA) and principal components analysis (PCA) to differentiate samples of olive oil from various sources and k-nearest neighbors (k-NN) algorithm to classify oils according to their quality. HCA with Ward's linkage was performed based on the Mahalanobis distance. To identify relevant clusters, 66% of the maximum distance between objects was used. Moreover, the concentration of seven phenols, which had the greatest impact on the result of a statistical analysis based on ANOVA, was used as input data to PCA. The PCA method made it possible to visualize the distinction between COO and EVOO samples. Classification accuracy and precision of supervised classification (k-NN) were validated using 10-fold cross-validation.

3. Results and Discussion

The results of the bioactive substance determination in extra virgin olive oils samples obtained from local Italian mills (EVOO) and from local distribution centers (COO) are summarized in Table 2. The results of EVOOs were divided into 11 Italian regions from which oils originated, while for COOs, 4 regions of origin were distinguished according to the information on the labels (see Table 1). Quantitative data obtained for Sicilian samples (EVOO and COO) were reported together in Table 2. In all the types of olive oil, 13 phenols (4 fat-soluble and 9 hydrophilic) were detected. The obtained results did not show any qualitative changes in the phenolic compound profile for olives of different geographical origin, as well as the process of their preparation.

According to the health claim for olive oil, authorized by European Union, a minimum of 250 mg kg^{-1} of selected phenols (hydroxytyrosol and its derivatives) is reported [27]. Figure 1A shows the content of these phenols in the studied olive oil samples. All the samples analyzed showed an average hydroxytyrosol, and its derivatives, content in accordance with the EU regulation [27]. As recommended by EFSA guidelines [6] a daily consumption between 10–20 g of EVOOs analyzed in this work ensures a daily intake of about 4–17 mg of oleuropein complex and hydroxytyrosol. It can also be seen that the nutraceutical value of COOs was lower than EVOOs, but more stable with smaller variations between samples, which may be related to technological factors such as the method of olive pressing or storage conditions. The only exceptions were the samples from Liguria, characterized by the lowest concentration of bioactive substances. Therefore, it is very important to control the quality of olive oil samples of both industrial and small-scale production.

During the research, in addition to phenolic alcohols and secoiridoids, phenolic acids and flavonoids were also determined; although they are not required for EU health claims, a proper assessment is needed since they contribute to increasing the health-promoting value of olive oil. Figure 1B shows the relative percentage of four chemical groups in EVOOs and COOs samples. It is relevant to note that the proportions of phenolic acids and flavonoids are similar in both oil groups. In the case of vitamin E, COOs, Sicilian and Ligurian oils were characterized by a higher relative content, while for secoiridoids it was the opposite. In both groups, the greatest fraction is represented by hydroxytyrosol and its derivatives (72–89%).

Tocochromanols or tocols belong to the group of tocopherols and tocotrienols described as vitamin E. These chemical compounds are fat-soluble antioxidants which inhibit lipid oxidation in various plant food products [28,29]. The major tocopherol found in olive oil is α-tocopherol, representing around 90% all isomers [30]. In this study, four chemical compounds, namely: α-, β-, γ-tocopherol and α-tocotrienol (their sum presented as vitamin E), were analyzed.

Table 2 summarize the tocopherol isomers and α-tocotrienol contents for EVOO and COO samples produced in Italy. The average vitamin E content in EVOOs and COOs was 169.0 ± 37.7 mg kg^{-1} and 151.7 ± 27.4 mg kg^{-1}, respectively. The major quantified isomer was α-tocopherol, representing 90.8% and 88.0% of total vitamin E respectively, in the range from 70.2 to 232.2 mg kg^{-1} of oil, followed by lower amounts of β (1.6–11.5 mg kg^{-1}) and γ isomers (1.3–17.7 mg kg^{-1}) as well as α-tocotrienol (0.9–5.8 mg kg^{-1}).

Table 2. Content (mg kg^{-1}) of bioactive molecules in olive oil samples analyzed divided into selected country/region.

		α-Tocopherol	α-Tocotrienol	α-Tocopherol	α-Tocopherol	Gallic Acid	Hydroxytyrosol	Tyrosol	Apigenin	Luteolin	Oleocanthal[a]	Oleacin[a]	Ligstroside Aglycone[b]	Oleuropein Aglycone[a]
Apulia	range	166.9–220.6	0.9–3.1	3.1–8.7	2.8–16.0	<LOD–6.5	22.8–146.5	25.5–71.3	0.1–11.7	5–24.8	<LOD–11.8	180.3–566.3	101.6–513.1	343.4–1248.9
	average	195.4	2.4	4.1	10.7	1.7	70.4	41.3	4.5	16.2	7.6	352.1	298.9	903.6
Sicily	range	80.4–184.0	1.3–3.7	2.9–9.2	2.7–15.5	1.7–19.7	27.5–80.8	5.4–79.5	<LOD–8.5	1.4–34.0	1.8–30.5	34.0–396.2	5.7–78.4	139.5–418.4
	average	123.7	1.8	5.5	5.4	7.6	62.8	37.3	2.0	16.1	8.0	148.6	44.3	274.6
Tuscany	range	113.1–232.2	1.1–4.9	1.6–7.1	3.4–17.1	<LOD–9.3	25.4–199.8	12.5–59.9	<LOD–14.0	7.6–53.0	<LOD–38.7	35.1–493.4	47.4–329.4	268.7–1514.0
	average	152.9	2.4	2.7	11.4	1.4	78.3	30.8	4.4	25.9	5.9	229.2	172.4	872.6
Lazio	range	164.8–211.4	3.3–4.4	2.8–8.6	6.1–15.2	<LOD–4.3	34.9–80.4	8.2–52.0	<LOD–19.5	3.5–26.9	0.1–17.0	71.4–331.4	11.0–141.0	89.5–1749.5
	average	190.3	3.9	5.4	10.9	1.3	59.4	27.5	7.2	16.3	4.7	197.0	81.8	830.7
Abruzzo	range	99.5–198.1	1.7–3.4	2.3–4.5	2.3–13.2	<LOD–1.7	50.5–117.3	8.9–34.2	<LOD–7.1	9.3–34.4	1.9–34.8	123.0–590.9	20.4–370.8	138.7–1785.9
	average	129.6	2.7	3.1	5.6	0.8	95.0	19.9	1.7	16.8	14.3	225.6	114.9	896.0
Campania	range	153.5–198.8	2.6–2.7	3.0–3.8	9.2–17.7	0.4–3.1	111.3–168.8	41.7–45.4	1.1–2.0	16.2–19.9	1.3–31.2	59.7–498.7	131.8–254.3	877.8–915.7
	average	176.2	2.6	3.4	13.4	1.7	140.0	43.5	1.5	18.1	16.2	279.2	193.0	896.8
Garda	range	89.6–123.9	2.1–2.6	1.8–2.0	3.8–6.3	0.2–3.1	68.3–76.7	33.2–54.8	0.5–8.0	21.0–40.9	5.3–8.8	156.4–290.7	86.7–130.5	438.3–1015.1
	average	112.6	2.3	1.9	5.1	1.5	70.9	40.2	3.2	30.6	7.3	239.1	107.7	799.6
Sardinia	range	113.4–139.6	2.6–3.0	2.1–3.2	4.7–8.3	<LOD–0.7	39.1–69.5	19.2–42.4	<LOD–4.2	13.5–32.2	3.8–8.9	154.6–272.4	110.7–288.0	590.8–1105.4
	average	130.5	2.8	2.6	5.7	0.2	55.0	30.3	1.5	20.8	6.2	213.1	165.4	778.0
Calabria	range	144.0–201.0	3.5–3.9	3.4–6.7	3.7–16.4	<LOD–0.2	56.7–145.3	25.3–52.4	6.0–15.4	20.8–32.8	0.4–14.8	54.7–408.2	86.0–169.0	562.7–1004.1
	average	163.7	3.7	5.1	10.4	0.1	105.2	38.2	9.9	28.7	7.7	287.3	116.6	777.9
Umbria	range	130.6–156.1	2.4–2.4	1.8–2.4	11.0–11.8	<LOD–0.4	47.3–52.8	21.2–24.4	6.2–6.9	26.7–36.9	2.6–4.7	137.0–254.0	81.8–97.8	666.5–892.2
	average	143.4	2.4	2.1	11.4	0.2	50.1	22.8	6.6	31.8	3.7	145.5	89.8	779.3
Liguria	range	70.2–125.9	2.1–2.9	3.5–5.3	4.3–4.7	3.2–5.1	122.2–129.3	37.2–37.8	1.1–2.3	21.0–23.7	2.4–2.7	28.2–32.0	39.5–45.3	102.6–143.0
	average	98.1	2.5	4.4	4.5	4.1	125.8	37.5	1.7	22.3	2.6	30.1	42.4	122.8
Italy	range	84.5–151.4	1.6–4.0	3.9–10.2	3.9–13.9	2.3–13.3	47.7–145.2	38.6–124.2	<LOD–3.5	11.4–26.4	0.3–15.1	23.2–259.2	21.1–136.5	211.1–692.3
	average	128.0	2.8	6.5	9.1	6.5	84.1	59.7	1.2	18.1	5.3	137.2	68.3	434.7
EU	range	90.1–209.0	1.5–4.3	3.9–11.5	6.8–17.5	0.8–11.0	37.8–208.1	21.5–90.7	<LOD–6.1	8.5–27.0	0.3–5.4	22.4–177.0	21.1–142.9	218.7–736.8
	average	137.9	3.0	8.4	10.8	5.7	78.4	52.4	0.9	17.5	2.0	104.8	45.6	378.0
Italy + EU	range	152.7–165.9	1.6–5.8	6.1–10.4	3.1–12.9	2.3–14.5	53.1–93.2	20.7–82.7	<LOD–4.4	19.2–22.9	0.7–2.7	64.4–94.9	27.1–59.8	116.7–619.9
	average	164.3	3.2	8.6	10.7	5.1	71.0	55.7	2.1	21.3	1.9	70.2	40.8	422.8

Bioactive molecules were quantitatively determined based on calibration curves obtained with the correspondent standard compound: [a] oleuropein, [b] verbascoside.

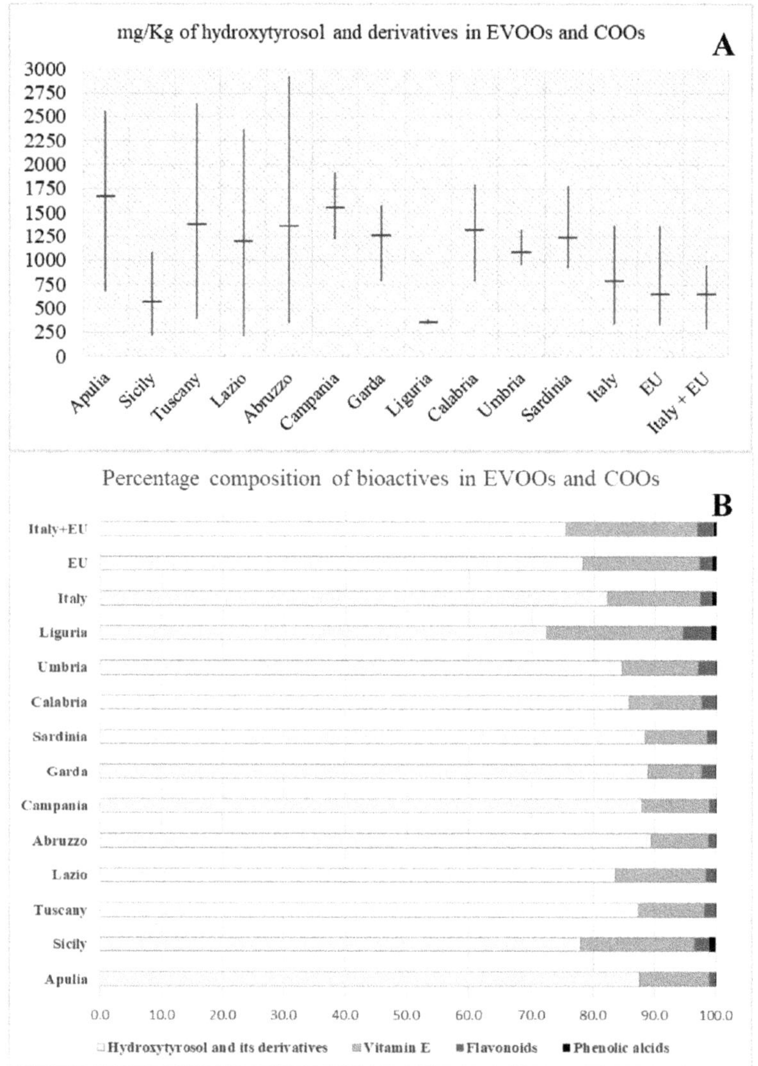

Figure 1. (**A**) mg kg^{-1} of hydroxytyrosol and its derivatives in EVOOs and COOs; (**B**) relative percentage of four chemical groups in EVOO and COO samples. For sample descriptions see Table 1. For Sicilian samples, both COO and EVOO were put together.

Comparable levels of tocopherols were mentioned by Saini et al. [28] and in USDA, ARS, FoodData Central database (https://fdc.nal.usda.gov/, accessed on 25 May 2020) [31]. The average content of α- and γ-tocopherol in olive oil was 143.5 and 8.3 mg kg^{-1}, respectively. It can be observed that some olive oil samples from different regions were characterized by lower values of these parameters. In the case of EVOOs, Liguria and Garda had a lower value of both parameters, while in the case of COOs, it should be noted that in samples designated as oils extracted from Sicilian and Italy + EU olives low content of γ-tocopherol was determined. In this study, no significant differences in the content of α-, β-, γ-tocopherol and α-tocotrienol between EVOO and COO were found ($p < 0.05$) (Table S1, Supplementary material). Therefore, tocopherols, tocotrienols and total vitamin E cannot be considered as a discriminant marker of olive oil quality, which is in agreement with Fiorini et al. who stated that the tocopherol concentration did not differ significantly between high- and low-quality Italian extra virgin olive oil [20].

Phenolic acids are a group of hydrophilic phenols in olive oil. They contribute to the color and organoleptic properties but also to the antioxidant and health properties of food [3]. Many external factors can affect the content of this chemical class in extra virgin olive oils, such as harvesting time, olive fruit ripeness, climatic conditions and oil extraction technologies [32]. For this reason, it was also decided to study this chemical class in EVOOs and COOs. Gallic acid was chosen as the representative of this phenolic group. The average phenolic acid content was 1.3 ± 1.7 mg kg^{-1} and 6.5 ± 3.7 mg kg^{-1} for EVOOs and COOs respectively. Gallic acid concentration was higher in commercial samples than in oils purchased from family farms. It should be noted that among EVOOs more than 90% of the samples contained less than 4 mg kg^{-1} of phenolic acids, and in the case of COOs, it was less than 20%. However, as much as 30% of COOs had a result higher than 10 mg kg^{-1} of gallic acid. The results for EVOOs were consistent with earlier literature data for monovarietal extra virgin olive oil samples [33]. However, the differences in average concentrations of gallic acid between EVOOs and COOs were not statistically significant (Table S1, Supplementary material), therefore the concentration of phenolic acids cannot be a factor differentiating the sample due to the technology of their extraction.

Tyrosol (Tyr) and hydroxytyrosol (HTyr) are the main phenolic alcohols found in olive oil. It should be mentioned that hydroxytyrosol is the first phytochemical compound approved by EFSA as exhibiting antioxidant properties and the resulting health benefits [6]. The total concentration of phenolic alcohols was from 35.3 to 216.9 mg kg^{-1} for EVOOs and from 70.7 to 296.0 mg kg^{-1} for COOs. The results showed that the concentrations of the target compounds differed in the range 22.8–199.8 mg kg^{-1} (Tyr) and 5.4–71.3 mg kg^{-1} (HTyr) for the studied EVOO samples, as well as in the range 37.8–208.1 mg kg^{-1} (Tyr) and 20.7–124.3 mg kg^{-1} (HTyr) for the COO samples. It should also be noted that all samples contained more tyrosol than hydroxytyrosol, which is consistent with reports from other authors [23,34,35]. The phenolic alcohols results obtained in this study for Apulian and Tuscanian EVOOs were within ranges which were determined in samples of monovarietal olive oils from Apulia and Tuscany by Bellumori et al. [35]. Furthermore, the results obtained for both EVOOs and COOs were comparable to those for high-quality Italian extra virgin olive oils [23].

Based on Table 2, it can be seen that tyrosol concentrations fluctuated at similar levels in both types of oil, obtained from low-scale mills and commercially available. In the case of hydroxytyrosol and phenolic alcohols, COOs samples were characterized by a higher content of these substances. However, no significant differences were found between the average values of tyrosol, hydroxytyrosol and the content of phenolic alcohols ($p < 0.05$) between EVOOs and COOs (Table S1, Supplementary material), which eliminates them as indicators enabling distinguishing of these samples. In general, the phenolic alcohol content in fresh oil is low, but increases during storage due to hydrolysis of secoiridoids [36]. Hence, a higher content of this chemical class in COOs may result, because commercial samples may have been stored for longer or even stored in inappropriate conditions, whereby the hydrolysis has been accelerated.

Flavonoids, which are the dominant secondary metabolites of plants, were another hydrophilic class of phenols found in olive oil. The main flavonoids identified in the olive oil were flavones present mostly in free form, luteolin and apigenin. Flavonoids are also characterized by many bioactive properties; therefore, they can have beneficial effects in the case of cardiovascular diseases, neurodegenerative disorders or cancer [37].

Two flavonoids, luteolin and apigenin, were determined in the tested samples (Table 2). The total average concentration of this chemical class was 26.6 ± 13.5 mg kg^{-1} for EVOOs and 18.9 ± 7.6 mg kg^{-1} for COOs. The most common flavonoid was luteolin, with values ranging from 1.2 mg kg^{-1} to 53.0 mg kg^{-1}. Apigenin concentration was lower than luteolin and determined values ranged from lower than LOQ to a maximum of 19.5 mg kg^{-1} for the EVOO sample from the Lazio region. According to Dugo et al. the content of luteolin in Italian high-quality extra virgin olive oil ranged from 4 to 149 mg kg^{-1}, whereas apigenin from <LOD to 38.8 mg kg^{-1} [24]. The results obtained in this study were within this range, however, the average concentrations were lower than those determined by Dugo et al. [24]. These results confirm that flavonoids as well as phenol concentration are influenced by cultivars, the geographical variability (like climate and territory) and the different production techniques.

It can be observed that the oils from Sicily, both marked as high quality and commercially available, were characterized by low flavonoids content. In the remaining groups, apigenin and luteolin values remained at similar levels. No significant differences were found in the luteolin, apigenin or flavonoid content between the EVOO and COO samples (Table S1, Supplementary material). The flavonoid concentration cannot, therefore, be considered as an indicator discriminating the quality of olive oil.

These results confirm prior reports that flavonoids are a chemical class constituting a small part of the phenol fraction but characterized by relative concentration stability. This statement is consistent with de Torres, who proved that flavonoid concentrations depend mainly on the degree of olive ripeness [37]. Fiorini et al. proved that HEVOO and LEVOO (high-price and low-price extra virgin olive oils) were characterized by similar values of marked flavonoids [20]. Bakhouche et al., however, proved that the concentration of luteolin and apigenin did not differ significantly between Spanish olive oils from different geographical areas [38].

Secoiridoids are phenols which are usually glycosidically bound and formed from secondary terpene metabolism. They are the most prevailing chemical class in olive oil. Ligstroside and oleuropein aglycones and their decarboxymethylated forms, namely oleocanthal (DCL) and oleacin (DCO) are the most common secoiridoids in olive oils. Oleuropein aglycone and oleacin are hydroxytyrosol esters, while the other two are tyrosol derivatives [39,40]. Following the European Union recommendation, secoiridoids are important compounds in the health claim for olive oil [27].

All four above secoiridoids were determined in this study (Table 2). The average concentrations of this chemical class were 1530 ± 548 mg kg^{-1} and 546 ± 182 mg kg^{-1}, for EVOOs and COOs, respectively. It should be emphasized that only one sample from the COO group contained more than 1000 mg kg^{-1} secoiridoids, while in the EVOO group as much as 65% of the samples showed this result. Except for one COO sample and two EVOO samples, the sum of aglycones was greater than the sum of DCO and DCL. Besides, hydroxytyrosol derivatives are more abundant than tyrosol derivatives in all cases, which is advantageous because hydroxytyrosol and its derivatives have higher antioxidant power than tyrosol [39]. Oleuropein aglycone was the most common secoiridoid in olive oil ranging from 39.5 to 1785.9 mg kg^{-1} for EVOOs and from 116.7 to 736.8 mg kg^{-1} for COOs. These results are in agreement with the outcomes of other authors using Italian and Spanish EVOOs [23,40].

Based on the analysis of variance and Tukey's test, statistically significant differences were observed for mean concentrations of oleuropein aglycone and secoiridoids of EVOO and COO samples ($p < 0.05$) (Table S1, Supplementary material). In addition to the differences between samples produced in low-scale mills and industrial plants, it was also achievable to distinguish samples from different geographical areas. Based on the concentration of secoiridoids, it was possible to find differences between EVOO samples from 11 regions of Italy and create three groups characterized by different contents of bioactive compounds. The group with the lowest secoiridoids concentration included samples from Sicily and Liguria, without significant differences between them ($p < 0.05$), but significant differences among other samples, while the samples with the highest concentration of these compounds came from Abruzzo, Apulia, Campania, Sardinia and Tuscany. On this basis, the content of secoiridoids can be considered as a potential indicator of the quality or geographical origin of olive oil. This may be due to the fact that the profile of secoiridoids is influenced not only by climatic but also by technological factors [39,41].

3.1. Multivariate Statistical Analysis

The main purpose of the chemometric analysis was to reveal specific relationships between oil samples or between variables to classify extra virgin olive oils by phenolic content and indicate which variables were discriminatory chemical indicators for EVOOs and COOs. As chemometric methods, hierarchical cluster analysis (HCA), principal component analysis (PCA) and k-nearest neighbors (k-NN) algorithm were used.

3.1.1. HCA (Hierarchical Cluster Analysis)

To understand the role of phenols as the quality indicators of extra virgin olive oils obtained from different places, HCA was used to group variables. Figure 2 presents a hierarchical dendrogram for 13 chemical substances. Based on the cluster analysis, it can be seen that phenols were grouped into three separate clusters based on the distance between data points in multidimensional space. In the case of reducing the number of variables, selecting variables from only one cluster should be avoided, as this may result in reduced model efficiency.

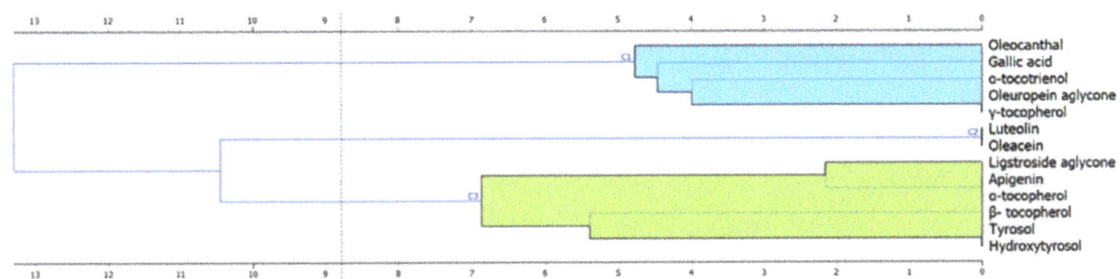

Figure 2. Hierarchical dendrogram for 13 chemical substances.

3.1.2. PCA (Principal Component Analysis)

The phenolic profile was also used as data for processing by PCA for more comprehensive studies focusing on the identification of potential quality markers of extra virgin olive oils of various origins. The input set consisted of the concentrations of seven selected phenols, namely: gallic acid, β-tocopherol, oleuropein aglycone, ligstroside aglycone, oleacein, hydroxytyrosol and apigenin, which were selected based on ANOVA. Selected compounds belonged to all three clusters presented in the dendrogram (Figure 2).

In Figure 3, bi-plot comparing scores along with loadings is depicted. The first two principal components explaining 63.1% of the total variance (45.2% and 17.9% for PC1 and PC2, respectively) divided the analyzed olive samples into three separate aggregations. The grouping of various extra virgin oils indicated that the qualitative and quantitative composition of phenolic compounds varies depending on both agronomic and technological factors. The division of samples into three subgroups was based on different contents of phenolic compounds. The first component (PC1) was positively correlated with chemical compounds from the class of secoiridoids and flavonoids. The other axis (PC2) was positively correlated with a representative of phenolic alcohols hydroxytyrosol. Interestingly, PC1 modelled the total polyphenol content with the richest oils on the right and the poorer ones on the left. This division allowed the separation of the first group containing only high-quality extra virgin olive oil samples. PC2 explained the difference between the second and third groups, namely, COO and EVOO, with the lowest content of phenol fraction (samples from Abruzzo, Lazio and Sicily). It can be observed that in the second group, apart from the COO samples, samples from Liguria were also grouped. The scatter plot with loadings described the behavior of variables, in other words, phenolic substances. It can be concluded that the levels of some compounds were positively correlated, for example, β-tocopherol and gallic acid, or oleuropein aglycone with ligstroside aglycone, with correlation coefficients higher than 0.5. Other substances were more negatively correlated, for example, apigenin with hydroxytyrosol or gallic acid.

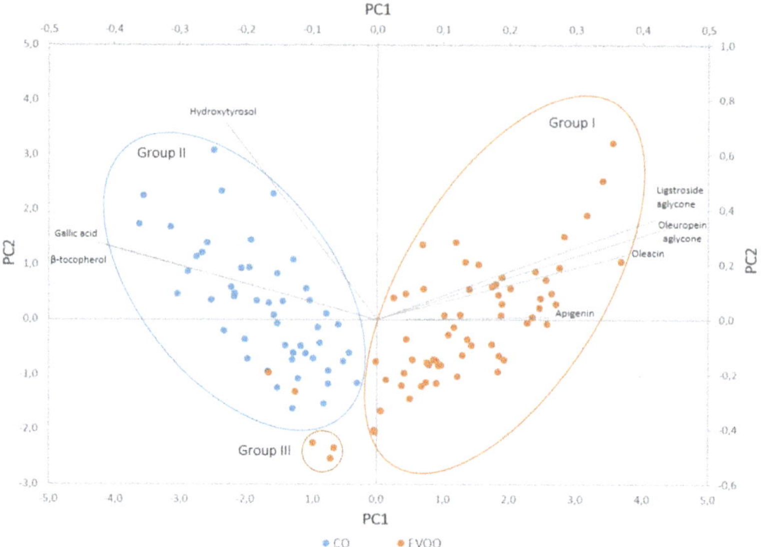

Figure 3. Bi-plot comparing scores along with loadings.

The PCA-biplot allows correlation between the selected bioactive compounds and the groups of objects (olive oil samples). For example, secoiridoids (ligstroside aglycone, oleuropein aglycone, oleacein) and flavonoids (as apigenin) were positively correlated with high-quality extra virgin olive oil samples. It was therefore proved, that high-quality oils were characterized by a higher concentration of secoiridoids and flavonoids. In turn, phenolic alcohols (as hydroxytyrosol), phenolic acids, gallic acid and β-tocopherol were associated with commercial olive oils and lower quality extra virgin olive oils.

3.1.3. k-NN (k-Nearest Neighbors)

To verify whether it is possible to assess the origin of extra virgin olive oil from small farms and supermarkets using concentrations of previously determined compounds as input variables for statistical analysis, k-NN was used to classify olive oils into two groups, namely EVOO and COO. EVOOs were the high-quality extra virgin olive oils samples from low-scale mills, while COOs were the lower-quality extra virgin olive oils which corresponded to commercially available extra virgin olive oils from industrial plants.

Input data were features previously selected for PCA. The data set was divided into two subsets, a training set and a test set. The accuracy of the classification model was measured using its recognition ability, in other words, the ability to classify samples from the training set, and the prediction ability, in other words, the ability to correctly classify samples from the test set. Through 10-fold cross-validation, it was found that the overall classification accuracy of the k-NN algorithm was 99%. The k-NN algorithm was then trained using 70% of the data selected at random to avoid bias and tested on the rest of the data. The use of k-NN resulted in an average of 94% correct classification with two false-negative results; this means that two samples of oils labelled as high-quality were classified as samples of lower quality. It should be noted that during multiple sampling, false-negative results were most common in samples from Liguria, Garda area or Sicily. During the PCA analysis, these samples were assigned to the second and third groups because they were characterized by a lower content of polyphenols compared to the samples from the first group (EVOO). It can, therefore, be concluded that by using the k-NN model, the quality of olive oil can be assessed, since samples classified as EVOO will have a higher nutraceutical potential compared to COO samples.

The model was also used to classify oil samples based on other features. According to Fiorini et al. [20], the classification was made applying the R parameter, in other words, the ratio of the concentration of tyrosol and hydroxytyrosol to the concentration of the above phenolic alcohols together with secoiridoid derivatives, the predictive efficiency was 85%. However, using HTyr and its derivatives, according to EFSA's health statement [6,26], the classification was 88%. Referring to point 3.2.5., secoiridoids were also used as input, resulting in 91% correct classification.

Based on the results of multivariate statistical analysis, it can be assessed that for the extra virgin olive oil quality index the most relevant approach was to select the concentrations of seven selected phenols, namely: gallic acid, β-tocopherol, oleuropein aglycone, ligstroside aglycone, oleacein, hydroxytyrosol and apigenin.

4. Conclusions

This work aimed to evaluate the concentration of potentially bioactive compounds in both commercial and locally-produced Italian extra virgin olive oils. Lipophilic and hydrophilic phenolic compounds were analyzed by HPLC coupled with fluorescence, photodiode array and mass spectrometry detection. According to the quantitative data attained, for both types of EVOO samples, the phenolic content exceeded the nutritional claims, reported in the EFSA Journal [5,6]. The results achieved highlight how the European standards guarantee an EVOO product of high quality; on the other hand, smaller olive oil producers manage to distinguish themselves in the production of EVOOs with a higher secoiriodoid and hydrotyrosol content. Extraction technology from olives was found as a parameter which could influence the chemical composition of EVOOs, and in particular the bioactive molecule content. So differences in antioxidant molecule composition not only depend on cultivar, quality, territory, climate, storage conditions and olive maturity index but also from production techniques.

The present study could pave the way for a future more extensive "comprehensive" work with a larger array of EVOO samples produced beyond the European community.

Supplementary Materials: The following are available online at http://www.mdpi.com/2304-8158/9/8/1120/s1, Table S1: Concentration of selected phenols in the extra virgin olive oils.

Author Contributions: Conceptualization, M.R. and F.C.; methodology, M.R. and F.C.; software, A.R.; validation, A.R. and F.S.; investigation, A.R. and F.S.; resources, P.D. and L.M.; data curation, A.R. and M.R.; writing—original draft preparation, A.R. and M.R.; writing—review and editing, M.R., F.C. and Ż.P.; supervision, M.R. and F.C.; project administration, P.D. All authors have read and agreed to the published version of the manuscript.

Funding: This research was funded by Cariplo Foundation within the "Agroalimentare e Ricerca" (AGER) program. Project AGER2-Rif.2016-0169, "Valorizzazione dei prodotti italiani derivanti dall'oliva attraverso Tecniche Analitiche Innovative-VIOLIN".

Acknowledgments: The authors are thankful to Shimadzu and Merck Life Science Corporations for the continuous support.

Conflicts of Interest: The authors declare no conflict of interest.

References

1. European Commission Website, News 4 February 2020. Available online: https://ec.europa.eu/info/news/producing-69-worlds-production-eu-largest-producer-olive-oil-2020-feb-04_en (accessed on 4 June 2020).
2. European Commission Website, Olive Oil, an Overview of the Production and Marketing of Olive Oil in the EU. Available online: https://ec.europa.eu/info/food-farming-fisheries/plants-and-plant-products/plant-products/olive-oil (accessed on 4 June 2020).
3. Bendini, A.; Cerretani, L.; Carrasco-Pancorbo, A.; Gómez-Caravaca, A.M.; Segura-Carretero, A.; Fernández-Gutiérrez, A.; Lercker, G. Phenolic molecules in virgin olive oils: A survey of their sensory properties, health effects, antioxidant activity and analytical methods. An overview of the last decade. *Molecules* **2007**, *12*, 1679–1719. [CrossRef] [PubMed]

4. Cioffi, G.; Pesca, M.S.; De Caprariis, P.; Braca, A.; Severino, L.; De Tommasi, N. Phenolic compounds in olive oil and olive pomace from Cilento (Campania, Italy) and their antioxidant activity. *Food Chem.* **2010**, *121*, 105–111. [CrossRef]

5. EFSA NDA Panel (EFSA Panel on Dietetic Products, Nutrition and Allergies). Scientific Opinion on Dietary Reference Values for Vitamin E as α-Tocopherol. *EFSA J.* **2015**, *13*, 4149. Available online: www.efsa.europa.eu/efsajournal (accessed on 25 May 2020). [CrossRef]

6. EFSA Panel on Dietetic Products, Nutrition and Allergies (NDA). Scientific Opinion on the substantiation of health claims related to polyphenols in olive and protection of LDL particles from oxidative damage (ID 1333, 1638, 1639, 1696, 2865), maintenance of normal blood HDL-cholesterol concentrations (ID 1639), maintenance of normal blood pressure (ID 3781), "anti-inflammatory properties" (ID 1882), "contributes to the upper respiratory tract health" (ID 3468), "can help to maintain a normal function of gastrointestinal tract" (3779), and "contributes to body defences against external agents" (ID 3467) pursuant to Article 13(1) of Regulation (EC) No 1924/2006. *EFSA J.* **2011**, *9*, 2033. Available online: www.efsa.europa.eu/efsajournal (accessed on 25 May 2020). [CrossRef]

7. Alves, F.C.G.B.S.; Coqueiro, A.; Março, P.H.; Valderrama, P. Evaluation of olive oils from the Mediterranean region by UV–Vis spectroscopy and Independent Component Analysis. *Food Chem.* **2019**, *273*, 124–129. [CrossRef]

8. Hosomi, A.; Arita, M.; Sato, Y.; Kiyose, C.; Ueda, T.; Igarashi, O.; Arai, H.; Inoue, K. Affinity for alpha-tocopherol transfer protein as a determinant of the biological activities of vitamin E analogs. *FEBS Lett.* **1997**, *409*, 105–108. [CrossRef]

9. Brigelius-Flohé, R.; Kelly, F.J.; Salonen, J.T.; Neuzil, J.; Zingg, J.M.; Azzi, A. The European perspective on vitamin E: Current knowledge and future research. *Am. J. Clin. Nutr.* **2002**, *76*, 703–716. [CrossRef] [PubMed]

10. Carrasco-Pancorbo, A.; Cerretani, L.; Bendini, A.; Segura-Carretero, A.; Gallina-Toschi, T.; Fernàndez-Gutirrez, A. Analytical determination of polyphenols in olive oils. *J. Sep. Sci.* **2005**, *28*, 837–858. [CrossRef]

11. Buettner, G.R. The pecking order of free radicals and antioxidants: Lipid peroxidation, alpha-tocopherol, and ascorbate. *Arch. Biochem. Biophys.* **1993**, *300*, 535–543. [CrossRef]

12. Sordini, B.; Veneziani, G.; Servili, M.; Esposto, S.; Selvaggini, R.; Lorefice, A.; Taticchi, A. A quanti-qualitative study of a phenolic extract as a natural antioxidant in the frying processes. *Food Chem.* **2019**, *279*, 426–434. [CrossRef] [PubMed]

13. Gómez-Rico, A.; Fregapane, G.; Desamparados Salvador, M. Effect of cultivar and ripening on minor components in Spanish olive fruits and their corresponding virgin olive oils. *Food Res. Int.* **2008**, *41*, 433–440. [CrossRef]

14. Capriotti, A.L.; Cavaliere, C.; Crescenzi, C.; Foglia, P.; Nescatelli, R.; Samperi, R.; Laganà, A. Comparison of extraction methods for the identification and quantification of polyphenols in virgin olive oil by ultra-HPLC-QToF mass spectrometry. *Food Chem.* **2014**, *158*, 392–400. [CrossRef]

15. Fanali, C.; Della Posta, S.; Dugo, L.; Russo, M.; Gentili, A.; Mondello, L.; De Gara, L. Application of deep eutectic solvents for the extraction of phenolic compounds from extra-virgin olive oil. *Electrophoresis* **2020**. [CrossRef]

16. Olmo-García, L.; Polari, J.J.; Li, X.; Bajoub, A.; Fernández-Gutiérrez, A.; Wang, S.C.; Carrasco-Pancorbo, A. Deep insight into the minor fraction of virgin olive oil by using LC-MS and GC-MS multi-class methodologies. *Food Chem.* **2018**, *261*, 184–193. [CrossRef]

17. Tsimidou, M.Z.; Nenadis, N.; Mastralexi, A.; Servili, M.; Butinar, B.; Vichi, S.; Winkelmann, O.; García-González, D.L.; Gallina Toschi, T. Toward a harmonized and standardized protocol for the determination of total hydroxytyrosol and tyrosol content in virgin olive oil (VOO). The pros of a fit for the purpose ultra high performance liquid chromatography (UHPLC) procedure. *Molecules* **2019**, *24*, 2429. [CrossRef]

18. Fanali, C.; Della Posta, S.; Vilmercati, A.; Dugo, L.; Russo, M.; Petitti, T.; Mondello, L.; De Gara, L. Extraction, analysis, and antioxidant activity evaluation of phenolic compounds in different Italian extra-virgin olive oils. *Molecules* **2018**, *23*, 3249. [CrossRef]

19. Klikarová, J.; Rotondo, A.; Cacciola, F.; Česlová, L.; Dugo, P.; Mondello, L.; Rigano, F. The phenolic fraction of Italian extra virgin olive oils: Elucidation through combined liquid chromatography and NMR approaches. *Food Anal. Methods* **2019**, *12*, 1759–1770. [CrossRef]

20. Fiorini, D.; Boarelli, M.C.; Conti, P.; Alfei, B.; Caprioli, G.; Ricciutelli, M.; Sagratini, G.; Fedeli, D.; Gabbianelli, R.; Pacetti, D. Chemical and sensory differences between high price and low price extra virgin olive oils. *Food Res. Int.* **2018**, *105*, 65–75. [CrossRef]

21. Ricciutelli, M.; Marconi, S.M.; Boarelli, C.; Caprioli, G.; Sagratini, G.; Ballini, R.; Fiorini, D. Olive oil polyphenols: A quantitative method by high-performance liquid-chromatography-diode-array detection for their determination and the assessment of the related health claim. *J. Chromatogr. A* **2017**, *1481*, 53–63. [CrossRef]

22. Piscopo, A.; Zappia, A.; De Bruno, A.; Poiana, M. Effect of the harvesting time on the quality of olive oils produced in Calabria. *Eur. J. Lipid Sci. Technol.* **2018**, *120*, 1700304. [CrossRef]

23. Klikarová, J.; Česlová, L.; Kalendová, P.; Dugo, P.; Mondello, L.; Cacciola, F. Evaluation of Italian extra virgin olive oils based on the phenolic compounds composition using multivariate statistical methods. *Eur. Food Res. Technol.* **2018**, *246*, 1241–1249. [CrossRef]

24. Dugo, L.; Russo, M.; Cacciola, F.; Mandolfino, F.; Salafia, F.; Vilmercati, A.; Fanali, C.; Casale, M.; De Gara, L.; Dugo, P.; et al. Determination of the Phenol and Tocopherol Content in Italian High-Quality Extra-Virgin Olive Oils by Using LC-MS and Multivariate Data Analysis. *Food Anal. Meth.* **2020**, *13*, 1027–1041. [CrossRef]

25. Russo, M.; Bonaccorsi, I.; Cacciola, F.; Dugo, L.; De Gara, L.; Dugo, P.; Mondello, L. Distribution of bioactives in entire mill chain from the drupe to the oil and wastes. *Nat. Prod. Res.* **2020**. [CrossRef]

26. Iqdiam, B.M.; Mostafa, H.; Goodrich-Schneider, R.; Baker, G.L.; Welt, B.; Marshall, M.R. High power ultrasound: Impact on olive paste temperature, malaxation time, extraction efficiency, and characteristics of extra virgin olive oil. *Food Bioprocess Tech.* **2018**, *11*, 634–644. [CrossRef]

27. European Commission. Commission Regulation (EU) No 432/2012 (16 May 2012) Establishing a list of permitted health claims made on foods, other than those referring to the reduction of disease risk and to children's development and health. *Off. J. Eur. Union* **2012**, *L136*, 1–40.

28. Saini, R.K.; Keum, Y.S. Tocopherols and tocotrienols in plants and their products: A review on methods of extraction, chromatographic separation, and detection. *Food Res. Int.* **2016**, *82*, 59–70. [CrossRef]

29. Tanno, R.; Kato, S.; Shimizu, N.; Ito, J.; Sato, S.; Ogura, Y.; Sakaino, M.; Sano, T.; Eitsuka, T.; Kuwahara, S.; et al. Analysis of oxidation products of α-tocopherol in extra virgin olive oil using liquid chromatography–tandem mass spectrometry. *Food Chem.* **2020**, *306*, 125582. [CrossRef]

30. Inarejos-García, A.M.; Santacatterina, M.; Salvador, M.D.; Fregapane, G.; Gómez-Alonso, S. PDO virgin olive oil quality-Minor components and organoleptic evaluation. *Food Res. Int.* **2010**, *43*, 2138–2146. [CrossRef]

31. Food Data Central. Available online: https://fdc.nal.usda.gov/ (accessed on 25 February 2020).

32. Tang, G.; Huang, Y.; Zhang, T.; Wang, Q.; Crommen, J.; Fillet, M.; Jiang, Z. Determination of phenolic acids in extra virgin olive oil using supercritical fluid chromatography coupled with single quadrupole mass spectrometry. *J. Pharm. Biomed. Anal.* **2018**, *157*, 217–225. [CrossRef]

33. Pancorbo, A.C.; Carretero, A.S.; Gutiérrez, A.F. Co-electroosmotic capillary eletrophoresis determination of phenolic acids in commercial olive oil. *J. Sep. Sci.* **2005**, *28*, 925–934. [CrossRef]

34. Pereira, C.; Costa Freitas, A.M.; Cabrita, M.J.; Garcia, R. Assessing tyrosol and hydroxytyrosol in Portuguese monovarietal olive oils: Revealing the nutraceutical potential by a combined spectroscopic and chromatographic techniques-based approach. *LWT* **2020**, *118*, 108797. [CrossRef]

35. Bellumori, M.; Cecchi, L.; Innocenti, M.; Clodoveo, M.L.; Corbo, F.; Mulinacci, N. The EFSA health claim on olive oil polyphenols: Acid hydrolysis validation and total hydroxytyrosol and tyrosol determination in Italian virgin olive oils. *Molecules* **2019**, *24*, 2179. [CrossRef]

36. Servili, M.; Esposto, S.; Fabiani, R.; Urbani, S.; Taticchi, A.; Mariucci, F.; Selvaggini, R.; Montedoro, G.F. Phenolic compounds in olive oil: Antioxidant, health and organoleptic activities according to their chemical structure. *Inflammopharmacology* **2009**, *17*, 76–84. [CrossRef]

37. De Torres, A.; Espínola, F.; Moya, M.; Alcalá, S.; Vidal, A.M.; Castro, E. Assessment of phenolic compounds in virgin olive oil by response surface methodology with particular focus on flavonoids and lignans. *LWT Food Sci. Technol.* **2018**, *90*, 22–30. [CrossRef]

38. Bakhouche, A.; Lozano-Sánchez, J.; Beltrán-Debón, R.; Joven, J.; Segura-Carretero, A.; Fernández-Gutiérrez, A. Phenolic characterization and geographical classification of commercial Arbequina extra-virgin olive oils produced in southern Catalonia. *Food Res. Int.* **2013**, *50*, 401–408. [CrossRef]

39. Amanpour, A.; Kelebek, H.; Selli, S. LC-DAD-ESI-MS/MS–based phenolic profiling and antioxidant activity in Turkish cv. Nizip Yaglik olive oils from different maturity olives. *J. Mass Spectrom.* **2019**, *54*, 227–238. [CrossRef]

40. Luque-Muñoz, A.; Tapia, R.; Haidour, A.; Justicia, J.; Cuerva, J.M. Direct determination of phenolic secoiridoids in olive oil by ultra-high performance liquid chromatography-triple quadruple mass spectrometry analysis. *Sci. Rep.* **2019**, *9*, 1–9. [CrossRef]

41. Pascale, R.; Bianco, G.; Cataldi, T.R.I.; Buchicchio, A.; Losito, I.; Altieri, G.; Genovese, F.; Tauriello, A.; Di Renzo, G.C.; Lafiosca, M.C. Investigation of the effects of virgin olive oil cleaning systems on the secoiridoid aglycone content using high performance liquid chromatography–mass spectrometry. *J. Am. Oil Chem. Soc.* **2018**, *95*, 665–671. [CrossRef]

Article

Optimization of Ultrasonicated Kaempferol Extraction from *Ocimum basilicum* Using a Box–Behnken Design and Its Densitometric Validation

Ammar B. Altemimi [1,*], **Muthanna J. Mohammed** [2], **Lee Yi-Chen** [3], **Dennis G. Watson** [3], **Naoufal Lakhssassi** [3], **Francesco Cacciola** [4,*] and **Salam A. Ibrahim** [5]

1   Department of Food Science, College of Agriculture, University of Basrah, Basrah 61004, Iraq
2   Department of Biology, College of Education for Pure Sciences, University of Mosul, Mosul 41002, Iraq; mjmk73@yahoo.com
3   School of Agricultural Sciences, Southern Illinois University at Carbondale, Carbondale, IL 62901, USA; yclee010689@gmail.com (L.Y.-C.); dwatson@siu.edu (D.G.W.); naoufal.lakhssassi@siu.edu (N.L.)
4   Department of Biomedical and Dental Sciences and Morphofunctional Imaging, University of Messina, 98125 Messina, Italy
5   Food and Nutritional Sciences Program, North Carolina A & T State University, Greensboro, NC 27411, USA; ibrah001@ncat.edu
*   Correspondence: ammaragr@siu.edu (A.B.A.); cacciolaf@unime.it (F.C.); Tel.: +964-773-5640-090 (A.B.A.); +39-090-676570 (F.C.)

Received: 2 September 2020; Accepted: 27 September 2020; Published: 29 September 2020

Abstract: Kaempferol (KA) is a natural flavonol that can be found in plants and plant-derived foods with a plethora of different pharmacological properties. In the current study, we developed an efficient extraction method for the isolation of KA from ultrasonicated basil leaves (*Ocimum basilicum*). We successfully employed a Box–Behnken design (BBD) in order to investigate the effect of different extraction variables including methanol concentration (40–80%), extraction temperature (40–60 °C), and extraction time (5–15 min). The quantification of KA yield was carried out by employing a validated densitometric high performance thin layer chromatography in connection with ultraviolet detection (HPTLC-VIS). The obtained data showed that the quadratic polynomial model (R^2 = 0.98) was the most appropriate. The optimized ultrasonic extraction yielded 94.7 ng/spot of KA when using methanol (79.99%) at 60 °C for 5 min. When using toluene-ethyl acetate-formic acid (70:30:1 $v/v/v$) as a solvent, KA was detected in basil leaves at an Retention factor (Rf) value of 0.26 at 330 nm. Notably, the analytical method was successfully validated with a linear regression of R^2 = 0.99, which reflected a good linear relationship. The developed HPTLC-VIS method in this study was precise, accurate, and robust due to the lower obtained results from both the percent relative standard deviation (%RSD) and SEM of the *O. basilicum*. The antioxidant activity of KA (half maximal inhibitory concentration (IC_{50}) = 0.68 µg/mL) was higher than that of the reference ascorbic acid (IC_{50} = 0.79 µg/mL) and butylated hydroxytoluene (BHT) (IC_{50} = 0.88 µg/mL). The development of economical and efficient techniques is very important for the extraction and quantification of important pharmaceutical compounds such as KA.

Keywords: kaempferol; *Ocimum basilicum*; high performance thin layer chromatography in connection with ultraviolet detection (HPTLC-VIS); box-Behnken design; optimization; validation

1. Introduction

Plants are sources of various phytochemicals such as terpenes, phenylpropanoids, diarylheptanoids, isothiocyanates, and sulfur compounds. These natural compounds are often explored for their potential use. For example, phytochemicals produced by herbs and spices are of interest due to their culinary and medicinal uses. Some of the phytochemical functions include antioxidant activities, the modulation of detoxification enzymes, the enhancement of the immune system, the reduction of inflammation, the modulation of steroid metabolism, antiviral and antibacterial effects, and the oxidative retardation of lipids [1,2]. Medicinal herbs can also be an alternative source of antioxidants outside of vitamin C, vitamin E, and carotenoids [3].

Basil (*Ocimum basilicum* L.) is an herbaceous annual plant that generally produces white-purple flowers and is one of the most important cultivated aromatic herbs in the world. The *Ocimum* genus belongs to the mint family (*Lamiaceae*), which contains dozens of medicinal plants that are grown for their high economic value. This genus is native to tropical and subtropical regions of Asia, Africa, and South America, and it consists of annual and perennial herbs and shrubs [4,5]. Basil can be cultivated in fields as well as under greenhouse conditions, which can result in different concentrations of chlorophylls and secondary metabolites. This herb is utilized for its medicinal values to treat headaches, coughs, diarrhea, constipation, warts, worms, and kidney malfunction [6]. Moreover, basil is an essential ingredient used in traditional culinary practices [7,8] and is known as rihan in Arabic [9]. Basil has many health benefits, and it is composed of essential nutrients such as vitamin A, vitamin C, calcium, phosphorus, and beta carotene [10]. It also contains flavonoids and phenolic compounds that act as reducing agents that contribute to its antioxidant activities [11]. Basil extract is commonly used as an herbal drug due to the many pharmacological effects such as anti-hyperglycemic [12], hypolipidemic, antiatherosclerotic [13], and anticancer activities [14]. Basil leaves contain several polyphenols, bioactive compounds [15], and phenolic acids (caffeic acid, caftaric acid, and rosmarinic acid) [16], and they are also rich in flavonoids such as anthocyanins, quercetin, kaempferol (KA), and luteolin. A very recent study conducted on the natural populations of basil indicated that it may have different chemical composition and biological activities [17]. Among flavonoids, KA is a natural flavonol antioxidant found in many fruits and vegetables. Several studies have focused on dietary KA due to its health benefits. In particular, it has been shown that KA may lower the risk of some chronic diseases, especially cancer [18]. In addition, KA has been reported to increase the body's antioxidant defense against free radicals, the common cause of cancer, resulting in an inverse relationship between the use of dietary KA and cancer [19,20]. Due to the high pharmacological benefits of basil, researchers worldwide are currently interested in developing simple, efficient, and reliable techniques to analyze different plant extracts for pharmaceutical purposes.

Plant extracts can be analyzed with Thin-Layer Chromatography (TLC) [21], Gas Chromatography-Mass Spectrometry (GC-MS) [22], and High-Performance Liquid Chromatography (HPLC) [23,24]. In recent years, the high performance TLC (HPTLC) densitometry method has grown in popularity because it is economic, sensitive, accurate, and reduces the time needed to process large samples. The increased sensitivity of this method and the ability to process many samples in short periods of time have led researchers to use TLC-densitometry over the popular HPLC [25]. Therefore, the aim of the present work was to optimize the ultrasonic-assisted extraction of KA from *Ocimum basilicum* by applying a Box–Behnken design (BBD) and to quantify such a flavonol with a simple, efficient, and validated HPTLC-VIS technique.

2. Materials and Methods

2.1. Sample and Sample Preparation

Samples of basil (*Ocimum basilicum*) were collected in Basrah, Iraq, and authenticated by the horticulture faculty at the College of Agriculture University of Basrah, Basrah, Iraq. Basil leaves were washed with distilled water, crushed in a blender, and then sealed in plastic bags and stored at −18 °C for two days before freeze-drying. Samples were shielded from light.

2.2. Solvent Mixture Screening

Different solvent mixtures (Table 1) were tested in order to maximize the extraction of total flavonoids. The higher flavonoid content positively reflected the amount of the desired KA compound. For each of the five solvent mixtures, 2 g of lyophilized basil were ground with a mortar and pestle with 20 mL of the appropriate solvent mixture. The mixture was transferred to a flask, and 100 mL of a different solvent mixture was added. The resulting solution was then transferred to a 200 mL cylindrical polypropylene container with a screw-on lid before insertion into an ultrasonic bath cleaner (UBC; Elmasonic P30, Elma Hans Schmidbauer GMBH, Singen, Germany) at a 37 kHz frequency, 35 W/cm^2 power, and 60 °C for 15 min.

Table 1. Solvent mixtures tested to maximize extraction of total flavonoids.

Solvent Mixture Code	Composition of Solvent Mixture
I	Methanol/water (50/50, *v/v*)
II	Acetic acid/acetone/water (10/60/30)
III	Methanol/acetic acid (90/10, *v/v*)
IV	Methanol/water/acetone (40/40/20, *v/v/v*)
V	Absolute methanol

2.3. Total Flavonoid Content

The total flavonoid content of each sample was quantified with the aluminum chloride colorimetric method [26]. Quercetin was used to make the standard calibration curve for total flavonoid content. Quercetin (5 mg) was dissolved in 1 mL of methanol, after which serial dilutions were prepared with methanol (5–200 mg/mL). Each of the serial dilutions was mixed with 0.6 mL of 2% aluminum chloride and incubated at room temperature for 60 min. A UV–Vis spectrophotometer (Sunny UV.7804C, Tokyo, Japan) was used to measure the absorbance of the reaction mixtures at the 420 nm wavelength and compared to a blank slide.

2.4. Ultrasonic-Assisted Extraction of KA

Lyophilized *Ocimum basilicum* (10 g) was weighed and placed in each one of the 200 mL glass flasks used. Different percentages of methanol (40%, 60%, and 80%) with 100 mL of solvent each were added to corresponding flasks and moved to a 120 mL cylindrical polypropylene container with a screw-on lid before insertion into the UBC. The UBC was operated at a 37 kHz frequency and 50% constant power for all treatments, and an adjustable water bath was used to reach the desired temperature before initiating ultrasonic treatment. According to the manufacturer's effective power rating, an ultrasonic power at a 50% power setting was used at 35 W/cm^2. The selected mixture was investigated at three different temperatures (40, 50, and 60 °C) over three ultrasonic treatment periods of 5, 10, and 15 min. Each UBC setting was repeated in triplicate with samples before changing to the next setting. After UBC treatment, the upper layer of samples was filtered (Whatman no. 1 paper) and subjected to rotary evaporation at 40 °C with a vacuum in order to remove the solvent [1].

2.5. Column Chromatography

Sequential purification through column chromatography was employed for partial purification for 14 different experiments with ultrasonicated *Ocimum basilicum* extracts. For the stationary phase, activated silica gel (pore size 60–120 mesh) was used, and for the mobile phase, a sequence of n-hexane, ethyl acetate, and petroleum ether was used. Additional column chromatography was completed after the crude methanol residue was partially dissolved in n-hexane and triturated with silica. Fractions were collected starting with n-hexane followed by ethyl acetate and petroleum ether. The collected fractions were then condensed. Due to their high fatty acid content, n-hexane and petroleum ether were discarded. A rotary evaporator was used to dry the collected ethyl acetate fractions before weighing. The result was yellow crystals of ethyl acetate fractions in test tubes. Each fraction was tested in order to confirm the presence of a single compound for ethyl acetate fractions using HPTLC.

2.6. Identification and Quantification of KA Using HPTLC-VIS

A Camag microliter syringe with a 0.22 μm syringe filter was used to filter the methanol extracts. Next, a Linomat 5 (Camag, Muttenz, Switzerland) was used to apply spots (5 μL) of methanol extracts (5 mg/mL) of *Ocimum basilicum* on a 20 × 10 cm sheet of silica gel 60 Fluorescent 254 (F_{254}) (E Merck, Darmstadt, Germany). The solvent solution was formed by toluene, ethyl acetate, and formic acid (70:30:1, *v/v/v*). The TLC plates were heated at 100 °C for 3 min and dipped into a reagent (2-aminoethyl diphenylborinate) before drying for 2 min in a cool air stream. The plate was subsequently dipped into a polyethylene glycol 400 reagent and allowed to dry for 5 min in a cool air stream. A Camag TLC scanner at 366 nm in absorbance mode was used for the densitometric scanning. A regression equation based on a calibration curve of the KA standard (Sigma Aldrich, St. Louis, MO, USA) was used to quantify KA concentrations.

2.7. Calibration Curve Preparation

KA in HPLC-grade methanol was prepared as a 1000 μg/mL stock solution. Various concentrations (i.e., 0.05, 0.1, 0.2, 0.3, and 0.4 μL) of the stock solution were spotted in order to achieve KA concentrations (50, 100, 200, 300, and 400 ng/spot) on three silica gel plates.

2.8. Response Surface Design

The presence of KA in the methanolic extracts of *Ocimum basilicum* was estimated from 14 fractions of column chromatography. The extraction process was optimized using a Box–Behnken response surface design (BBD) with three factors at three levels. The three factors were methanol concentration (X_1, %), extraction temperature (X_2, °C), and extraction time (X_3, min). The factor levels are shown in Table 2. The extraction yield of KA was the response variable. Design-Expert version 12 software was implemented using the following second-order polynomial model to describe the effect of the factors and levels:

$$Y = b_0 + \sum_{i=1}^{3} b_i X_i + \sum_{i=1}^{3} b_{ii} X^2_i + \sum_{i \neq j=1}^{3} b_{ii} X_i X_j \tag{1}$$

where Y is the predicted response b_0 is the intercept; b_1, b_2, and b_3 are the linear coefficients of methanol concentration (X_1), extraction temperature (X_2), and extraction time (X_3), respectively; b_{11}, b_{22}, and b_{33} are the squared coefficient of methanol concentration, extraction temperature, and extraction time, respectively; and b_{12}, b_{13}, and b_{23} are the interaction coefficients of methanol concentration, temperature of sonication, and extraction time, respectively. The factor settings were represented as $X_i–X_j$.

Table 2. Extraction variables selected for *Ocimum basilicum* optimization. KA: kaempferol.

Independent Variable	Ranges of Independent Variable			Dependent Variable	Goal
	−1	**0**	**+1**		
Methanol concentration (%)	40	60	80		
Extraction temperature (°C)	40	50	60	KA Yield	Maximized
Extraction time (min)	5	10	15		

2.9. Validation Method

Several steps were implemented in order to validate the HPTLC method. The range of compound concentrations was measured for linearity, with results expressed as a correlation coefficient from linear regression. A sample was prepared by adding a known concentration of compounds to a previously quantified sample. This sample was analyzed at three different times on the same day in order to assess intra-day precision and daily for 5 consecutive days in order to quantify inter-day precision. Systemic error was calculated in order to estimate the accuracy of the analytical method. The relative standard deviation of the linearity data and a calibration curve slope were used to calculate the limit of detection (LOD) and the limit of quantitation (LOQ) for KA.

2.10. Free Radical Scavenging Activity

Methanol extracts and KA were evaluated for antioxidant activities using the 2,2-diphenyl-1-picryl-hydrazyl-hydrate (DPPH) radical scavenging method described by Mensor et al. [27], with slight modifications. Dimethyl Sulfoxide DMSO (SIGMA) was used to dissolve test samples and mixed with 20 mg/mL of a DPPH methanol solution in order to yield concentrations of 10, 50, 100, 500, and 1000 µg/mL. After all samples were allowed to sit at room temperature for 30 min, absorbance values were measured at 517 nm and converted into the percentage of antioxidant activity as follows:

$$\% \text{ Inhibition} = (\text{Absorbance of control} - \text{Absorbance of test sample}) \times 100/\text{Absorbance of control}$$

Butylated hydroxytoluene (BHT) and L-ascorbic acid were used as reference standards with the unit of probits for the inhibition ratio. Probit values were plotted against the logarithmic values of test samples concentrations using a linear regression curve to establish the half maximal inhibitory concentration (IC_{50}) values (µg/mL). All analyses were repeated in triplicate, with results expressed as mean ± standard deviation (SD) and compared using the Waller–Duncan test with alpha = 0.05.

2.11. Surface Method Model Response and Validity Testing

Experiment results were analyzed with the Design-Expert™ (version 12) software (alpha = 0.05). The three factors of methanol concentration, extraction temperature, and extraction time were simultaneously optimized using the response surface method (RSM). A subsequent ultrasonic-assisted extraction experiment was completed in triplicate using the optimized conditions, and the KA yield was compared with the predicted values for the validation of the model. A mean comparison using Tukey's test was performed with STATISTICA 13 (alpha = 0.05).

3. Results and Discussion

3.1. Screening of Total Flavonoid Content in Basil Leaves

The total flavonoid content from basil leaves was determined in this study. The solvent mixtures used for the extraction contained different ratios of methanol, water, acetone, and acetic acid (Table 1). The different solvent mixtures used for the extraction significantly ($p < 0.05$) affected the total flavonoid content. The methanol/water (50/50, *v/v*) solvent extraction provided the highest yield, whereas the methanol/acetic acid (90/10, *v/v*) solvent extraction had the lowest flavonoid content when performing extraction on freeze-dried materials (Figure 1). Solvent polarity played a critical role in the extraction of total flavonoid content, and the methanol/water mixture led to the best performance acting as a great solvent system for polar antioxidants [28,29].

Several studies have been conducted to develop efficient procedures for the large-scale industrial application of natural antioxidants. Due to higher yields, ethanol and methanol are commonly used as solvents to extract flavonoids [30]. However, in a study conducted by Kobus-Cisowska et al. [31] on *Morus alba* fruits, acetone-based mixtures turned out to be more effective than methanol-based mixtures for phenolic and flavonoid extractions. In order to optimize and maximize extraction procedures, it is critical for researchers to take the food matrices and the concentration of the solvent into consideration when extracting phenolic compounds and flavonoids from herbal plants [32].

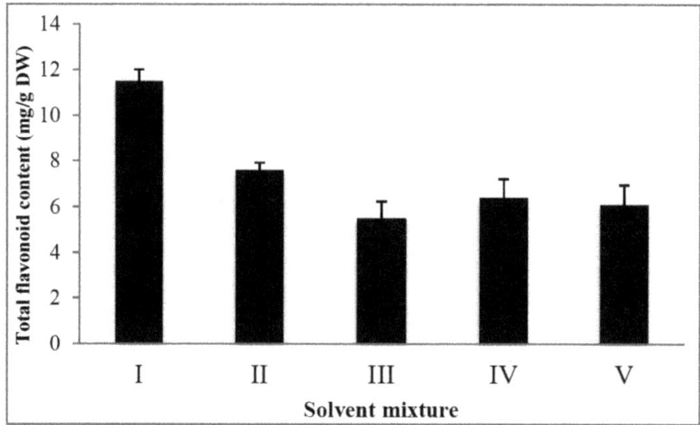

Figure 1. Type of solvent mixture used for extraction.

3.2. HPTLC-VIS Analysis of KA

From the screening study, it was shown that the methanol–water solvent exhibited the highest yield of flavonoids from basil leaves when compared to the other four solvents. As a consequence, this type of extraction solvent was selected for the KA extraction. The ultrasonication extraction technique was performed on the extraction of the KA-containing methanol extract. To separate the bioactive compounds, the ultrasonicated *Ocimum basilicum* extracts from 15 experiments were partially purified with sequential purification through column chromatography. The results obtained from the column chromatography showed that among the hexane, ethyl acetate, and petroleum ether extractions, only ethyl acetate impacted the presence of KA in the extract. The ethyl acetate extraction was subsequently selected for TLC analysis. The test tubes were collected and the mobile phase was evaporated in order to obtain the pure, yellow crystals from the ethyl acetate fractions. The KA band in the ultrasonicated *Ocimum basilicum* extracts was verified by being compared with a standard of the KA sample using TLC (Figure 2). A validated HPTLC-VIS technique was used to quantify the KA for all of the experimental Box–Behnken design (BBD) runs. A solvent system consisting of toluene-ethyl acetate-formic acid (70:30:1 *v/v/v*) was selected, and this system provided a sharp and well-defined KA peak at an Retention factor (Rf) value of 0.26 (Figure 3). To our knowledge, this is the first report of the solvent system used for the isolation of KA. The clear resolution exhibited on the chromatography demonstrates that this solvent system is capable of efficiently separating KA from the methanol extract of the ultrasonicated *Ocimum basilicum* (Figure 4).

Sample concentrations within the range of 50–400 ng/spot were selected to study the range of linearity (Table 3). The obtained data were close to unity and fit a linear model ($R^2 = 0.99$). Validation parameters such as the LOD and the LOQ were used to confirm the accuracy and precision of the proposed method with reproducible results. The validation parameters in the current study were large (LOD = 8.16 ng; LOQ = 18.142 ng), indicating that the method used here was more sensitive than the HPTLC methods used for previous KA analyses [33]. A recovery exceeding 98% was observed for the accuracy study (Table 4). For the precision test, a recovery of less than 2% was observed for the relative standard deviation (RSD) for inter-day and inter-day variations, suggesting that the proposed method had outstanding precision (Table 5).

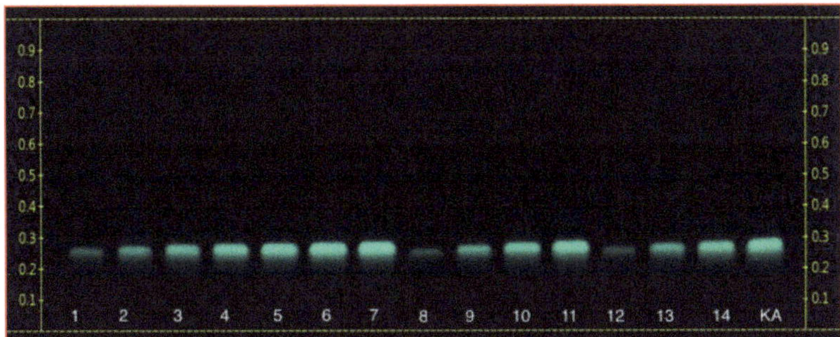

Figure 2. Photograph of thin layer chromatography (TLC) of ethyl acetate fraction at 366 nm. Kaempferol (KA); spots: 1–14 for ultrasonicated *Ocimum basilicum* extracts.

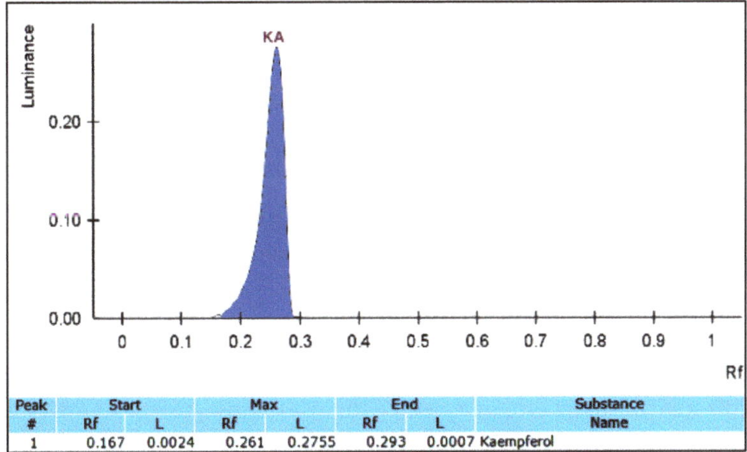

| Peak | Start | | Max | | End | | Substance |
#	Rf	L	Rf	L	Rf	L	Name
1	0.167	0.0024	0.261	0.2755	0.293	0.0007	Kaempferol

Figure 3. High performance thin layer chromatography (HPTLC) chromatogram of the KA standard. Rf, Retention factor. #, number; L, lowest distance.

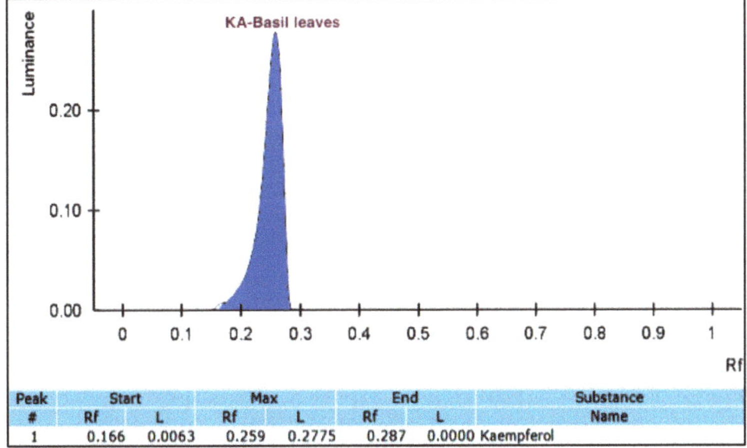

| Peak | Start | | Max | | End | | Substance |
#	Rf	L	Rf	L	Rf	L	Name
1	0.166	0.0063	0.259	0.2775	0.287	0.0000	Kaempferol

Figure 4. HPTLC chromatogram of KA in the methanol extract of ultrasonicated basil leaves (*Ocimum basilicum*) under the optimized conditions. Rf, Retention factor. #, number; L, lowest distance.

Table 3. Validation parameters for High-Performance Liquid Chromatography (HPLC). LOD: limit of detection; LOQ: limit of quantitation. Rf, Retention factor.

Validation Parameter	Value
Linearity range	(50–400) ng/spot
Correlation coefficient	0.99
LOD (ng)	8.16
LOQ (ng)	18.142
Specificity	Specific
Rf value	0.261

Table 4. Data of accuracy for KA.

Concentration (ng/spot)		Amount of KA Found (Mean)	SD	%RSD	%Recovery ($n = 3$)
Taken	Added				
150	0	148.90	0.59	0.401	99.26
150	25	171.88	0.71	0.413	98.21
150	50	197	0.57	0.289	98.50
150	75	222	0.66	0.297	98.66
150	100	247	0.43	0.174	98.80

SD = standard deviation; RSD = relative standard deviation.

Table 5. Data for inter-day and intra-day precision for KA.

Concentration (µg/spot)	Inter-Day Precision ($n = 5$)			Intra-Day Precision ($n = 3$)		
	Peak Area (Mean)	SD	%RSD	Peak Area (Mean)	SD	%RSD
5	875	9.77	1.11	922.11	1.14	0.12
10	1172.23	9.98	0.85	1290.88	0.94	0.07
15	1388.16	1.49	0.10	1465	7.77	0.53
20	1642.99	1.13	0.06	1805.29	3.31	0.18
25	1923.23	1.09	0.05	2218.22	3.03	0.13

SD = standard deviation; RSD = relative standard deviation.

3.3. Model Fitting

HPTLC-VIS was used to quantify the KA yield from each experimental BBD run (Table 6). A regression analysis was performed, and four proposed models were analyzed. The quadratic model appeared to be the best fit (adjusted $R^2 = 0.9138$), followed by the cubic model (adjusted $R^2 = 0.9033$), 2 Factor Interaction (FI) model (adjusted $R^2 = 0.5462$), and linear model (adjusted $R^2 = 0.1986$) (Table 7). The adjusted R^2 of the quadratic model was close to 1, implying a certain degree of correlation between the observed and predicted values. The adequate precision was calculated in order to determine the signal-to-noise ratio, and in this study, the adequate precision reading was 17.58, which was larger than the desired value (>4.0). Such a large value would be indicative of an adequate signal and could be used to navigate through the design space. The analysis of variance (ANOVA) of the fitted quadratic polynomial model for the KA yield is presented in Table 8. The lack of a fit F-value test for the model described the deviation in the data around the fitted model. The obtained value was not significant (*p*-value = 0.265), indicating that the result might have been due to pure error, which therefore validated the RSM results. In other words, if the model did not adequately fit the data, the resultant lack-of-fit value would be significant. The optimization of a fitted response surface is likely to provide false results [34]. The model in this study was significant (*p*-value = 0.082). The achieved results were in agreement with previous published data on *Ocimum sanctum* where an HPTLC method was developed involving a Box–Behnken-supported design for the simultaneous optimization, validation, and quantification of polyphenols in an aqueous alcoholic extract [35].

Table 6. Box–Behnken design (BBD) matrix for the optimization of extraction of KA yield (ng/spot).

Run	Methanol Concentration (A) (%)	Extraction Temperature (B) (°C)	Extraction Time (C) (min)	KA (ng/spot)
1	60	50	10	90
2	80	60	10	95
3	80	50	5	91
4	60	60	5	91.1
5	40	40	10	91.2
6	40	60	10	90.2
7	40	50	15	90
8	80	40	10	89
9	60	50	10	90.3
10	40	50	5	90.6
11	60	60	15	90.2
12	60	40	15	89
13	80	50	15	90.5
14	60	40	5	87

Table 7. Regression analysis results.

	Model F Value	R^2	Adjusted R^2	Predicted R^2
	Linear	0.3835	0.1986	−0.3353
KA Yield	2FI	0.7557	0.5462	−0.3151
	Cubic	0.8924	0.9033	-
	Quadratic	0.9735	0.9138	0.9088

FI (factor interaction).

Table 8. Analysis of variance (ANOVA) results for KA extraction.

Source	Sum of Square	Degree of Freedom	Mean Square F	F-Value	Prob > F
Model	37.55	9	4.17	16.32	0.0082
Residual	1.02	4	0.2556	-	-
Lack of fit	0.9775	3	0.3258	7.24	0.2651
Pure error	0.0450	1	0.0450	-	-

3.4. Effect of Extraction Parameters on KA Yield of Ultrasonicated Ocimum Basilicum and RSM Analysis

The ANOVA and the contributions of each independent variable for the extraction of KA from ultrasonicated *Ocimum basilicum* were determined. The linear, interaction, and quadratic variables were significant (*p* < 0.05), thus indicating that the KA yield was affected by all of the variables, the interactions between variables, and the square of each variable. The model in the study showed a high degree of precision (R^2 = 0.9735), and the experiment values were reliable (% Coefficient of

Variation (CV) = 0.55) [36]. Aiming to visualize the relationship between the independent variables and the KA yield, three-dimensional (3D) plots were constructed based on the generated quadratic polynomial model equation of coded factors:

$$\text{KA yield} = 90.15 + 0.4375 \times A + 1.2875 \times B + 1.73543 \times 10^{-15} \times C + 1.75 \times AB + 0.025 \times$$
$$AC + -0.725 \times BC + 1.2 \times A^2 + 1.89196 \times 10^{-14} \times B^2 + -0.825 \times C^2$$

The parameter variables in this study were methanol concentration (A), extraction temperature (B), and extraction time (C). The effects of the parameter variables and their interactions on the KA contents in the ultrasonicated *Ocimum basilicum* were examined. The third variable was assigned to be constant at the intermediate setting, while the two independent variables were displayed on three-dimensional surface plots.

Figure 5A shows that the yield of the KA content rapidly increased and reached the maximum value at 0 level of extraction time when the methanol concentration increased from 40% to 79.99%. The extraction temperature was fixed at 60.02 °C. Figure 5B shows how the KA content was affected by the interaction between the methanol concentration and the extraction time at a fixed extraction temperature level of 0. The maximum KA content was obtained when the methanol concentration was at 80%. The KA yield content decreased slightly when the methanol concentration was increased to 80%. The extraction time was fixed at 5.0 min. Figure 5C shows that the KA yield rapidly increased and reached the maximum value at 0 level of methanol concentration when the extraction temperature increased from 40 to 60 °C. The extraction time was fixed at 5.0 min. Interestingly, when the extraction temperature exceeded 60 °C, the yield of the extracted KA rapidly decreased. This result confirmed the fact that the solubility of the solute was enhanced when the temperature increased, resulting in a higher yield. However, the solvent density could be simultaneously reduced when the temperature increased, consequently decreasing the total flavonoid yield. This would mean that either a positive or negative effect could occur when the temperature increased [37]. This finding is in agreement with the study conducted by Peng et al. [38] in which the thermal degradation of flavonoids and the yield of KA decreased when the number of acoustic cavitation bubbles decreased.

3.5. Optimization and Verification of the Model for Extraction Parameters

Design-Expert[TM] software (version 12) was used to determine the optimized extraction process parameter. The desirability function was applied to the optimizing stage in order to maximize the yield of KA. The optimal conditions for the three parameters of the extraction process were estimated: The methanol concentration was 79.99%, the extraction temperature was 60.02 °C, the extraction time was 5.0 min, and the KA yield was 94.7 ng/spot. The extraction process was repeated, the optimum extraction conditions, i.e., an 80% methanol concentration at 60 °C for 5 min, and the KA yield was 94.65 ± 0.5. There was no significant difference ($p > 0.05$) between the predicted value and the experiment value, which demonstrated that the model proposed in this study can be utilized for extracting KA from ultrasonicated *Ocimum basilicum*.

3.6. RSM Validation

The KA yield evaluation results were found to be within the limits of the yield checkpoint. Further validation was performed on the RSM results, and the percentage prediction error was found to be 0.789% when comparing the experiment values of the response with the anticipated values. This provided sufficient evidence to establish the validity of the generated equation and to describe the domain of applicability of the RSM model. The goodness of fit was excellent for the predicted and experiment value linear correlation plot ($R^2 = 0.9905$; $p < 0.05$) (Figure 6). Overall, there was a high prognostic ability due to the low magnitudes of error and the significant values of R^2.

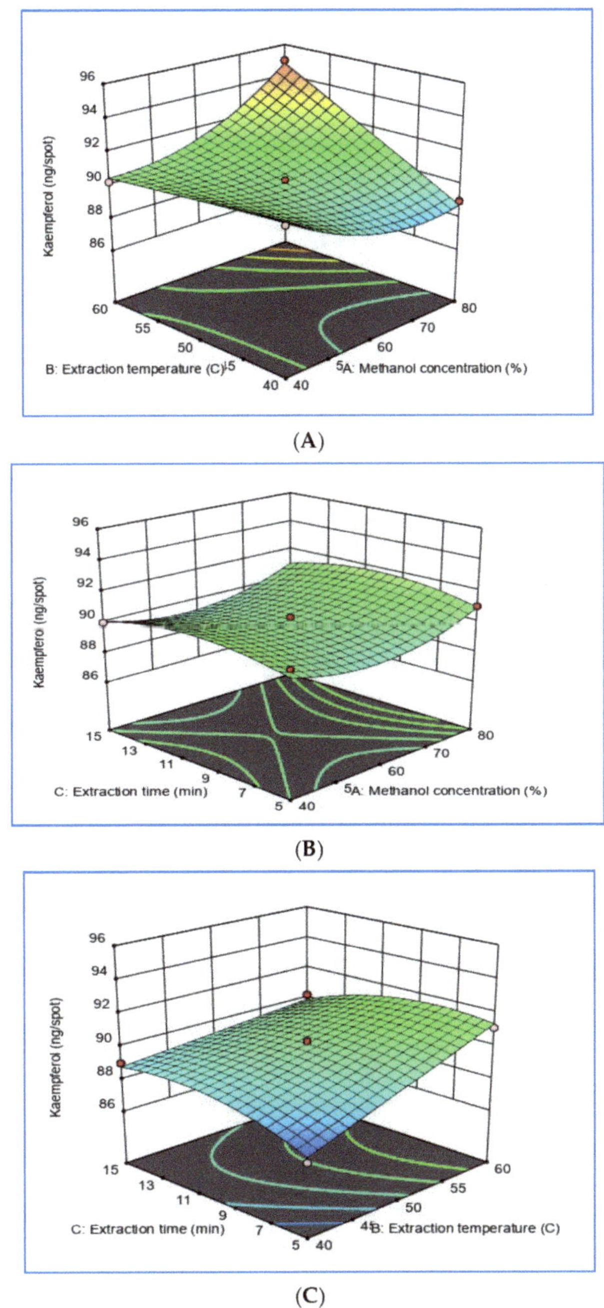

(A)

(B)

(C)

Figure 5. Response surface model 3D plots showing the effects of independent variables of ultrasonicated *Ocimum basilicum* on KA yield. (**A**) Methanol concentration and extraction temperature, (**B**) methanol concentration and extraction time, and (**C**) extraction temperature and extraction time.

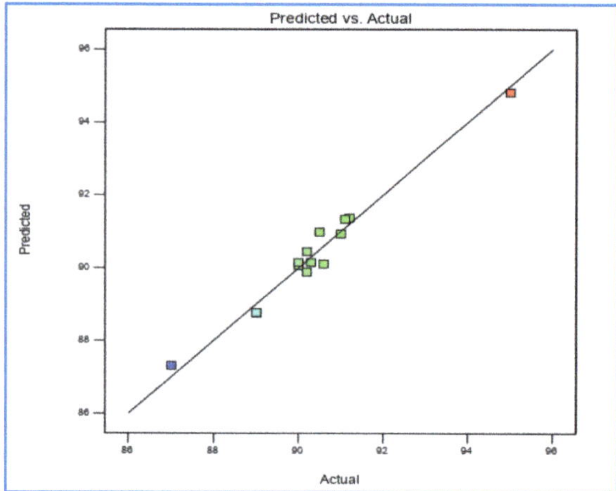

Figure 6. Linear correlation plot between actual and predicted values for KA yield.

3.7. Antioxidant Activity

Data on the antioxidant activities of the methanol extract of ultrasonicated basil leaves (*Ocimum basilicum*) under the optimized conditions, as well as their isolated compound (KA), are presented in Table 9. The antioxidant activity of the methanol extract of ultrasonicated *Ocimum basilicum* was $IC_{50} =$ 50.10 µg/mL. In contrast, the antioxidant activity of KA ($IC_{50} = 0.68$ µg/mL) was higher with respect to the reference ascorbic acid ($IC_{50} = 0.79$ µg/mL) and BHT ($IC_{50} = 0.88$ µg/mL). The present study was in accordance with the work of Shafique et al. [39], who reported that *O. basilicum* extracts exhibited and offered higher antioxidant activity compared to synthetic antioxidant BHT.

Table 9. Inhibition concentrations of test samples scavenging 50% of the 2,2-diphenyl-1-picryl-hydrazyl-hydrate (DPPH) radical.

Test Samples	IC_{50} (µg/mL)
Methanol extract of ultrasonicated *Ocimum basilicum*	50.10 ± 0.513
Isolated KA	0.68 ± 0.021 [c]
L-ascorbic acid	0.79 ± 0.015 [b]
BHT	0.88 ± 0.026 [a]

Half maximal inhibitory concentration (IC_{50}). [a], [b], [c] Means with different superscripts are significantly different at $p \leq 0.05$.

4. Conclusions

The data obtained from the current study showed that the HPTLC-VIS densitometric method and the BBD were efficient for the identification and quantitative analysis of KA from basil leaves. The interaction and the quadratic terms of each factor from all three variables had significant effects on the KA yield. A quadratic model for the KA yield was derived with $R^2 = 0.99$. The current study revealed that the use of methanol (79.99%) at 60 °C for 5 min comprised the optimal conditions for extracting a high yield of KA from ultrasonicated basil leaf extracts. Under these conditions, the KA yield was 94.7 (ng/spot), which was consistent with the predicted yield value. The isolated KA exhibited and offered a higher antioxidant activity compared to the reference of ascorbic acid and BHT. This study thus demonstrated, for the first time, the utilization of a solvent system that can be used with higher efficiency during the HPTLC-VIS analysis of KA. The used solvent system provided a high quality resolution of KA peaks. The LOD and LOQ values were found to be comparatively low, thus supporting the high sensitivity of the developed method. In addition, this method was supported by statistical analysis that highlighted how the proposed method is reproducible and specific for this type of extraction.

Author Contributions: Methodology, M.J.M.; investigation, A.B.A.; writing—original draft preparation, D.G.W.; writing—review and editing, N.L., F.C., and S.A.I.; supervision, L.Y.-C. All authors have read and agreed to the published version of the manuscript.

Funding: This research received no external funding.

Acknowledgments: The authors would like to thank the Food Science Department in the College of Agriculture, University of Basrah, Basrah, Iraq for providing facilities and equipment.

Conflicts of Interest: The authors declare no conflict of interest.

References

1. Altemimi, A.; Lightfoot, D.A.; Kinsel, M.; Watson, D.G. Employing response surface methodology for the optimization of ultrasound assisted extraction of lutein and β-carotene from spinach. *Molecules* **2015**, *20*, 6611–6625. [CrossRef] [PubMed]

2. Mathivha, P.L.; Msagati, T.A.; Thibane, V.S.; Mudau, F.N. Phytochemical Analysis of Herbal Teas and Their Potential Health, and Food Safety Benefits: A Review. In *Herbal Medicine in India*; Springer: Singapore, 2020; pp. 281–301.

3. Al-Maskari, M.Y.; Hanif, M.A.; Al-Maskri, A.Y.; Al-Adawi, S. Basil: A natural source of antioxidants and neutraceuticals. In *Natural Products and Their Active Compounds on Disease Prevention*; Nova Science Publishers, Inc.: Hauppauge, NY, USA, 2012; pp. 463–471.

4. Abd El-Azim, M.H.; Abdelgawad, A.A.; El-Gerby, M.; Ali, S.; El-Mesallamy, A.M. Phenolic compounds and cytotoxic activities of methanol extract of basil (*Ocimum basilicum* L.). *J. Microb. Biochem. Technol.* **2015**, *7*, 182–185. [CrossRef]

5. Jahan, R.; Jannat, K.; Shoma, J.F.; Khan, M.A.; Shekhar, H.U.; Rahmatullah, M. Drug Discovery and Herbal Drug Development: A Special Focus on the Anti-diarrheal Plants of Bangladesh. In *Herbal Medicine in India*; Springer: Singapore, 2020; pp. 363–400.

6. Pratama, H.N.; Dwiyanti, R.D.; Muhlisin, A. Growth of Malassezia furfur in Media with The Addition of Basil (*Ocimum basilicum* Linn) Powder. *Trop. Med. Int. Health* **2020**, *2*, 26–33. [CrossRef]

7. Agarwal, C.; Sharma, N.L.; Gaurav, S.S. An analysis of basil (*Ocimum* sp.) to study the morphological variability. *Ind. J. Fund. Appl. Life Sci.* **2013**, *3*, 521–525.

8. Skrypnik, L.; Novikova, A.; Tokupova, E. Improvement of Phenolic Compounds, Essential Oil Content and Antioxidant Properties of Sweet Basil (*Ocimum basilicum* L.) Depending on Type and Concentration of Selenium Application. *Plants* **2019**, *8*, 458. [CrossRef] [PubMed]

9. Bilal, A.; Jahan, N.; Ahmed, A.; Bilal, S.N.; Habib, S.; Hajra, S. Phytochemical and pharmacological studies on *Ocimum basilicum* L.: A review. *Int. J. Curr. Res. Rev.* **2012**, *4*, 73–83.

10. Pedro, A.C.; Moreira, F.; Granato, D.; Rosso, N.D. Extraction of bioactive compounds and free radical scavenging activity of purple basil (*Ocimum basilicum* L.) leaf extracts as affected by temperature and time. *An. Acad. Bras. Ciênc.* **2016**, *88*, 1055–1068. [CrossRef]

11. Prinsi, B.; Morgutti, S.; Negrini, N.; Faoro, F.; Espen, L. Insight into Composition of Bioactive Phenolic Compounds in Leaves and Flowers of Green and Purple Basil. *Plants* **2020**, *9*, 22. [CrossRef]

12. Abdelrahman, N.; El-Banna, R.; Arafa, M.M.; Hady, M.M. Hypoglycemic efficacy of *Rosmarinus officinalis* and/or *Ocimum basilicum* leaves powder as a promising clinico-nutritional management tool for diabetes mellitus in Rottweiler dogs. *Vet. World* **2020**, *13*, 73–79. [CrossRef]

13. Zhan, Y.; An, X.; Wang, S.; Sun, M.; Zhou, H. Basil polysaccharides: A review on extraction, bioactivities and pharmacological applications. *Bioorg. Med. Chem.* **2020**, *28*, 115179. [CrossRef]

14. Aburjai, T.A.; Mansi, K.; Azzam, H.; Alqudah, D.A.; Alshaer, W.; Abuirjei, M. Chemical Compositions and Anticancer Potential of Essential Oil from Greenhouse-cultivated *Ocimum basilicum* Leaves. *Ind. J. Pharm. Sci.* **2020**, *82*, 179–184. [CrossRef]

15. Nazir, M.; Ullah, M.A.; Younas, M.; Siddiquah, A.; Shah, M.; Giglioli-Guivarc'h, N.; Abbasi, B.H. Light-mediated biosynthesis of phenylpropanoid metabolites and antioxidant potential in callus cultures of purple basil (*Ocimum basilicum* L. var purpurascens). *Plant Cell Tissue Organ Cult.* **2020**, *142*, 107–120. [CrossRef]
16. Da Silva, L.A.L.; Pezzini, B.R.; Soares, L. Spectrophotometric determination of the total flavonoid content in *Ocimum basilicum* L. (*Lamiaceae*) leaves. *Pharmacogn. Mag.* **2015**, *11*, 96. [PubMed]
17. Elansary, H.O.; Szopa, A.; Kubica, P.; Ekiert, H.; El-Ansary, D.O.; Al-Mana, F.A.; Mahmoud, E.A. Saudi *Rosmarinus officinalis* and *Ocimum basilicum* L. Polyphenols and Biological Activities. *Processes* **2020**, *8*, 446. [CrossRef]
18. Chen, C.H. Dietary Inducers of Detoxification Enzymes. In *Xenobiotic Metabolic Enzymes: Bioactivation and Antioxidant Defense*; Springer: Cham, Switzerland, 2020; pp. 221–234.
19. Garro-Aguilar, Y.; Cayssials, V.; Achaintre, D.; Boeing, H.; Mancini, F.R.; Mahamat-Saleh, Y.; Karakatsani, A. Correlations between urinary concentrations and dietary intakes of flavonols in the European Prospective Investigation into Cancer and Nutrition (EPIC) study. *Eur. J. Nutr.* **2020**, *59*, 1481–1492. [CrossRef] [PubMed]
20. Paswan, S.K.; Srivastava, S.; Rao, C.V. Wound healing, antimicrobial and antioxidant efficacy of *Amaranthus spinosus* ethanolic extract on rats. *Biocatal. Agric. Biotechnol.* **2020**, *26*, 101624. [CrossRef]
21. Legerská, B.; Chmelová, D.; Ondrejovič, M. TLC-Bioautography as a fast and cheap screening method for the detection of α-chymotrypsin inhibitors in crude plant extracts. *J. Biotechnol.* **2020**, *313*, 11–17. [CrossRef]
22. Sagbo, I.J.; Orock, A.E.; Kola, E.; Otang-Mbeng, W. Phytochemical screening and gas chromatography-mass spectrometry analysis of ethanol extract of *Scambiosa columbabria* L. *Pharmacogn. Res.* **2020**, *12*, 35. [CrossRef]
23. Al-Samman, A.M.M.A.; Siddique, N.A. Gas Chromatography-Mass Spectrometry (GC-MS/MS) Analysis, Ultrasonic Assisted Extraction, Antibacterial and Antifungal Activity of *Emblica officinalis* Fruit Extract. *Pharmacogn. J.* **2019**, *11*, 315–323. [CrossRef]
24. Doctor, N.; Kayan, B.; Parker, G.; Vang, K.; Yang, Y. Stability and extraction of vanillin and coumarin under subcritical water conditions. *Molecules* **2020**, *25*, 1061. [CrossRef]
25. Bhardwaj, P.; Banarjee, A.; Jindal, D.; Kaur, C.; Singh, G.; Kumar, P.; Kumar, R. Validation of TLC-densitometry method for estimation of catechin in acacia catechu heartwood. *Pharm. Chem. J.* **2020**, *54*, 184–189. [CrossRef]
26. Marinova, D.; Ribarova, F.; Atanassova, M. Total phenolics and total flavonoids in Bulgarian fruits and vegetables. *J. Univ. Chem. Technol. Metall.* **2005**, *40*, 255–260.
27. Mensor, L.L.; Menezes, F.S.; Leitão, G.G.; Reis, A.S.; Santos, T.C.D.; Coube, C.S.; Leitão, S.G. Screening of Brazilian plant extracts for antioxidant activity by the use of DPPH free radical method. *Phytother. Res.* **2001**, *15*, 127–130. [CrossRef] [PubMed]
28. Naczk, M.; Shahidi, F. Phenolics in cereals, fruits and vegetables: Occurrence, extraction and analysis. *J. Pharm. Biomed. Anal.* **2006**, *41*, 1523–1542. [CrossRef] [PubMed]
29. Alothman, M.; Bhat, R.; Karim, A.A. Antioxidant capacity and phenolic content of selected tropical fruits from Malaysia, extracted with different solvents. *Food Chem.* **2009**, *115*, 785–788. [CrossRef]
30. Chávez-González, M.L.; Sepúlveda, L.; Verma, D.K.; Luna-García, H.A.; Rodríguez-Durán, L.V.; Ilina, A.; Aguilar, C.N. Conventional and Emerging Extraction Processes of Flavonoids. *Processes* **2020**, *8*, 434. [CrossRef]
31. Kobus-Cisowska, J.; Szczepaniak, O.; Szymanowska-Powałowska, D.; Piechocka, J.; Szulc, P.; Dziedziński, M. Antioxidant potential of various solvent extract from *Morus alba* fruits and its major polyphenols composition. *Ciênc. Rural* **2020**, *50*. [CrossRef]
32. Vural, N.; Cavuldak, Ö.A.; Akay, M.A.; Anlı, R.E. Determination of the various extraction solvent effects on polyphenolic profile and antioxidant activities of selected tea samples by chemometric approach. *J. Food Meas. Charact.* **2020**, *14*, 1286–1305. [CrossRef]
33. Siddiqui, N.; Aeri, V. Optimization of betulinic acid extraction from *Tecomella undulata* bark using a box-Behnken design and its densitometric validation. *Molecules* **2016**, *21*, 393. [CrossRef]
34. Altemimi, A.; Lakhssassi, N.; Baharlouei, A.; Watson, D.G.; Lightfoot, D.A. Phytochemicals: Extraction, isolation, and identification of bioactive compounds from plant extracts. *Plants* **2017**, *6*, 42. [CrossRef]
35. Ilyas, U.K.; Deepshikha, P.H.; Vidhu, A. Densitometric Validation and Optimisation of Polyphenols in *Ocimum sanctum* Linn by High Performance Thin-layer Chromatography. *Phytochem. Anal.* **2015**, *26*, 237–246.
36. Chadha, H. Application of Response Surface Methodology for Combination of Herbal Extracts Against Antioxidant and Antipsoriatic Activity. *Asian J. Pharm.* **2020**, *14*, 169–174.

37. Li, W.; Liu, Z.; Wang, Z.; Chen, L.; Sun, Y.; Hou, J.; Zheng, Y. Application of accelerated solvent extraction to the investigation of saikosaponins from the roots of *Bupleurum falcatum*. *J. Sep. Sci.* **2010**, *33*, 1870–1876. [CrossRef] [PubMed]

38. Peng, L.-X.; Zou, L.; Zhao, J.; Xiang, D.-B.; Zhu, P.; Zhao, G. Response surface modeling and optimization of ultrasound-assisted extraction of three flavonoids from tartary buckwheat (*Fagopyrum tataricum*). *Pharmacogn. Mag.* **2013**, *9*, 210. [PubMed]

39. Shafique, M.; Khan, S.J.; Khan, N.H. Study of antioxidant and antimicrobial activity of sweet basil (*Ocimum basilicum*) essential oil. *Pharmacologyonline* **2011**, *1*, 105–111.

 *foods*

Article

Isolation of Microalgae from Mediterranean Seawater and Production of Lipids in the Cultivated Species

Imane Haoujar [1,2,*], Nadia Skali Senhaji [1], Francesco Cacciola [3,*], Hicham Chairi [4], Manuel Manchado [5], Jamal Abrini [1], Mohammed Haoujar [6], Kamal Chebbaki [2], Marianna Oteri [7], Francesca Rigano [8], Domenica Mangraviti [8], Luigi Mondello [8,9,10,11] and Adil Essafi [12]

[1] Laboratory of Biotechnology and Applied Microbiology, Department of Biology, Faculty of Sciences of Tetouan, Abdelmalek Essaadi University, 93000 Tetouan, Morocco; abrinij@hotmail.com (J.A.); senhajin@hotmail.com (N.S.S.)

[2] Specialized Center in Zootechnics and Marine Aquaculture Engineering, National Institute of Fisheries Research, 93200 M'Diq, Morocco; chebbakikamal@yahoo.fr

[3] Department of Biomedical, Dental, Morphological and Functional Imaging Sciences, University of Messina, 98125 Messina, Italy

[4] Laboratory of Biological Engineering, Agri-Food and Aquaculture, Department of Biology, Faculty poly-disciplinary of Larache, Abdelmalek Essaadi University, 92000 Larache, Morocco; hicham.chairi@yahoo.fr

[5] IFAPA El Toruño, 11500 El Puerto de Santa Maria, Spain; manuel.manchado@juntadeandalucia.es

[6] IAV Hassan II, Madinat Al Irfane 6202 Instituts, 10101 Rabat, Morocco; haoujartopographie@gmail.com

[7] Department of Veterinary Sciences, University of Messina, Viale Annunziata, 98168 Messina, Italy; moteri@unime.it

[8] Department of Chemical, Biological, Pharmaceutical and Environmental Sciences, University of Messina, 98168 Messina, Italy; frigano@unime.it (F.R.); dmangraviti@unime.it (D.M.); lmondello@unime.it (L.M.)

[9] Chromaleonts.r.l., c/o Department of Chemical, Biological, Pharmaceutical and Environmental Sciences, University of Messina, 98168 Messina, Italy; lmondello@unime.it

[10] Department of Sciences and Technologies for Human and Environment, University Campus Bio-Medico of Rome, 00128 Rome, Italy; lmondello@unime.it

[11] BeSeps.r.l., c/o Department of Chemical, Biological, Pharmaceutical and Environmental Sciences, University of Messina, 98168 Messina, Italy; lmondello@unime.it

[12] National Aquaculture Development Agency, 10100 Rabat, Morocco; adil.essafi@yahoo.com

*   Correspondence: imane.haoujar2015@gmail.com (I.H.); cacciolaf@unime.it (F.C.);Tel.: +21-262-422-8056 (I.H.); +39-090-676-6570 (F.C.)

Received: 7 October 2020; Accepted: 30 October 2020; Published: 4 November 2020

Abstract: Isolation and identification of novel microalgae strains with high lipid productivity is one of the most important research topics to have emerged recently. However, practical production processes will probably require the use of local strains adapted to commanding climatic conditions. The present manuscript describes the isolation of 96 microalgae strains from seawater located in Bay M'diq, Morocco. Four strains were identified using the 18S rDNA and morphological identification through microscopic examination. The biomass and lipid productivity were compared and showed good results for *Nannochloris* sp., (15.93 mg/L/day). The lipid content in the four species, namely *Nannochloropsis gaditana*, *Nannochloris* sp., *Phaedactylum tricornutum* and *Tetraselmis suecica*, was carried out by HPLC-MS highlighting the identification of up to 77compounds.

Keywords: microalgae; isolation; identification; lipid productivity; HPLC-MS

1. Introduction

Aquatic microalgae are photosynthetic microorganisms that live with a variety of other species and meet different ecological requirements [1]. They represent one of the first forms of life on earth, and they have been found in oceans for more than 3 billion years since terrestrial environmental components were installed [2,3]. Fifty thousand microalgae species with diverse groups like Cyanobacteria, prokaryotic and eukaryotic microalgae have been discovered in oceans and freshwater lakes, ponds, and rivers around the world, however, only thirty thousand of them have been analyzed [1]. Thanks to their biological property, microalgae can be used as a new source of compounds in several biotechnological applications, including wastewater treatment [4], biodiesel production [5], and as supplements for human and animal dietary [6,7].

A large amount of funding has recently been invested to select the best species of microalgae with high bioactive metabolites [8]. Microalgae represent several sources of bioactive compounds, such as polyphenols, carotenoids, polysaccharides, omega-3, fatty acids, and polyunsaturated fatty acids (PUFA) [9–12]. The lipid concentrations in microalgae are between 20% and 70%, and the fatty acid composition in algal cells is highly dependent on genetic and phenotypic agents, including environmental and culture conditions [13]. Large scale lipid production will command the use of competitive species that are easy to grow and adapt to local environmental conditions. Isolating strains of microalgae with rapid growth, high intrinsic lipid content, and high biomass productivities is a primary necessity [14–16]. The quantity of total lipids in the form of glycolipid, phospholipid, and neutral lipid is varied considerably among and within groups of microalgae [6]. Many prior studies have identified the percent of omega-3 fatty acids between 30 and 40% of their total fatty acids in several species of microalgae like *Nannochloropsis* sp. (EPA), *Schizochytrium limacinum* (DHA), and *Phaeodactylum tricornutum* [17].

As a consequence, microalgae have great potential in the human diet as supplements for the treatment of physiological aberrations, prevention management, and used as synthetic dietary supplements to providing sustainable natural resources [18].

The isolation and identification of algal species from a natural environment is a well-established procedure. Each species has different growing conditions, including several regulations and key physic-chemical factors controlling growth and development such as temperature, pH, salinity, and silicate for diatoms [19,20].

The main objective of the present study was to evaluate the effect of the dilution technique and agar plate to separate and isolate different microalgae from Bay M'diq, Morocco. Biomass productivity, lipid productivity, and lipids content of some isolated microalgae were evaluated. Moreover, the 18SrDNA encoding gene of the microalgal isolates was sequenced to confirm the identification of the isolates. The ultra-high performance liquid chromatography coupled to mass spectrometry (UHPLC-MS) has been successfully used for the analysis of several lipid classes such as triacylglycerols (TGs), glycolipids, and phospholipids [21].Ahigh chromatographic resolution was achieved by serially coupling two narrow-bore partially porous columns [22]. A reversed-phase stationary phase, viz.octadecylsilica (C18), was used in order to obtain a good separation of lipids, according to their different hydrophobicity. As for MS detection, a single quadrupole was used as an analyzer and the atmospheric pressure chemical ionization (APCI) interface was used as an ion source. The APCI, different from the most common electrospray (ESI), is able to generate some in-source fragmentation, useful for structure elucidation, especially when a single quadrupole is employed as the MS analyzer and tandem MS experiments cannot be carried out [23–25].

2. Materials and Methods

2.1. Sampling and Isolation

Different species of microalgae were collected from seawater at Bay of M'diq, which is located in the north-western part of Morocco, between Sebta (35°54′ N, 5°1′10″W) and Capo Negro (35°40′ N,

16′40″ W). Samples were selected from three collection sites: (i) proximity to a fish farm of the bay; (ii) off the coast of Martil, and (iii) at Kabila Port (Figure 1). A volume of 1 L of seawater sample was taken using clean bottles at a depth of 0.5 m and then stored in cool boxes for transportation to the laboratory.

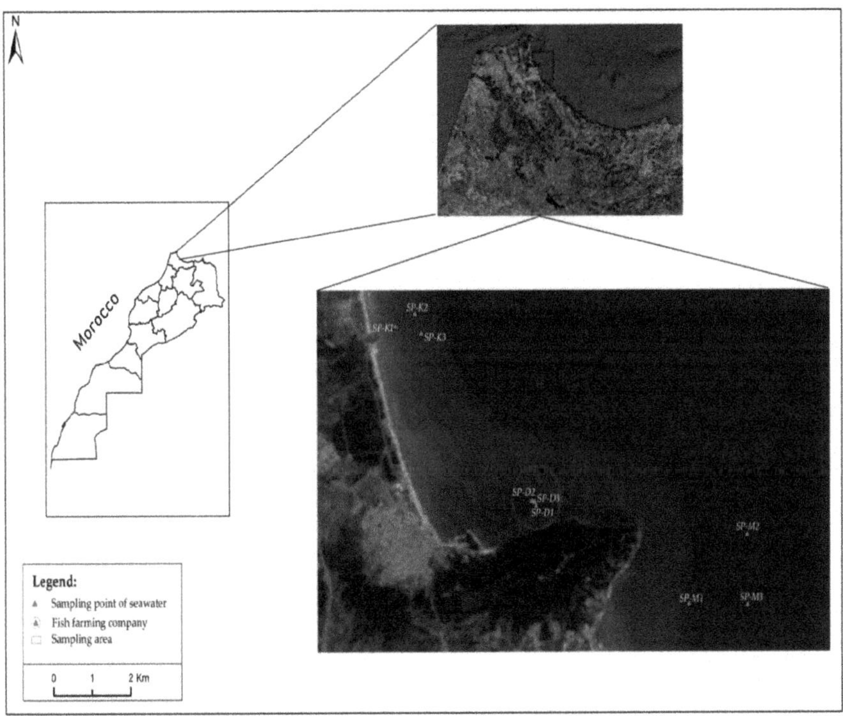

Figure 1. Distribution map of sampling points in M'diq Bay, Morocco.

Seawater samples were taken from December 2017 to February 2018 in a regular manner with constant frequencies with nine samples per month.

Once in the laboratory, the samples are purified using two methods; they were inoculated on an agar plate and in a liquid medium (Guillard F/2) that contained: $NaNO_3$ 8.82×10^{-4} M; $NaH_2PO_4 \cdot H_2O$ 3.62×10^{-5} M; $Na_2SiO_3 \cdot 9H_2O$ 1.06×10^{-4} M; $FeCl_3 \cdot 6H_2O$ 1.17×10^{-5} M; $Na_2EDTA \cdot 2H_2O$ 1.17×10^{-5} M; $MnCl_2 \cdot 4H_2O$ 9.10×10^{-7} M; $ZnSO_4 \cdot 7H_2O$ 7.65×10^{-8} M; $CoCl_2 \cdot 6H_2O$ 4.20×10^{-8} M; $CuSO_4 \cdot 5H_2O$ 3.93×10^{-8} M; $Na_2MoO_4 \cdot 2H_2O$ 2.60×10^{-8} M; Thiamin HCl (Vit. B1) 2.96×10^{-7} M; Biotin (Vit. H) 2.05×10^{-9} M; and Cyanocobalamin (Vit. B12) 3.69×10^{-10} M [26].

For those isolated on an agar plate, the first group of samples was filtered through a series of membranes of decreasing mesh (33, 20, and 0.45 μm). Each membrane was directly plated on agar plates containing F/2medium solidified with 1.5% of agar and incubated in a light chamber at two temperatures (20 and 26 °C). After growth, different colonies were transferred to tubes of 8 mL [27]. However, the second group of samples was isolated by successive dilutions, starting with 1 mL of sample to 10 mL of F/2 medium in order to transfer only one cell into a test tube, thereby establishing a single-cell isolate [28,29]. This procedure of dilution was repeated with serial dilutions from 10^{-2} to 10^{-10} until obtained unicellular tubes medium [30,31] (Figure 2).

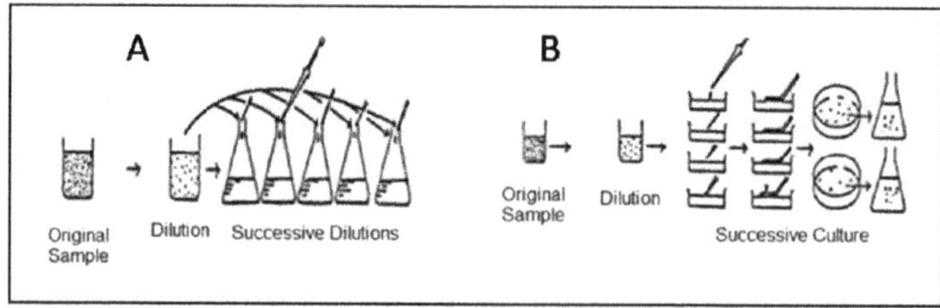

Figure 2. Method of cell isolation with (**A**) successive dilution, and (**B**) inoculation on an agar plate.

During the isolation process, the F/2 medium using in this experiment was divided into two groups; one with and the other one without silicate, to select only strains that required this nutrient. All strains incubated at two temperatures (20 and 26 °C) to evaluate the effect of temperature on cell growth; All cultivations were alimented with atmospheric CO_2. The light was provided by warm white fluorescent bulbs at 25 W/m^2 and operated on a light/dark cycle of 12/12 h.

After cell growth, all tubes were inoculated by transferring to 125 mL flasks add by 70 mL of F/2 medium, then incubated with a photoperiod of 18/6 h.

2.2. Strain Identification

The morphological identification of different isolated strains was carried out by microscopic observation. At the end, for molecular identification, DNA was extracted using 1030 mg of microalgal dried biomass.

Samples were homogenized using the FastDNA kit for 40 s at speed setting 5 in the Fastprep FG120 instrument (Bio101, Inc., Vista, CA, USA) but using the reagents of the ISOLATE II Genomic DNA Kit (Bioline) kit following the manufacture's protocols. DNA was quantified spectrophotometrically using the Nanodrop ND8000.

PCR amplifications were carried out using the Platinum Multiplex Master Mix (Thermofisher) in a 25 µL final volume containing 20 ng DNA, 300 nM each of specific forward and reverse primers, and 12.5 µL of Enzyme Premix (Thermofisher). The amplification protocol used was as follows: initial 11 min denaturation and enzyme activation at 95 °C, 30 cycles of 20 s at 95 °C, 1 min at 56 °C (60 °C) and 2 min at 72 °C and final extension step of 10 min at 72 °C. The 18S rDNA primers used were 18SF and euk516R [32,33] and for plastidic gene ribulose-1,5-bisphosphate carboxylase/oxygenase large subunit (rbcl) TetraRBCL_F and R [34] (Table 1). The PCR products were electrophoresed on a 2% agarose gel after staining with ethidium bromide and visualized via ultraviolet transillumination. Following the PCR reaction, free dNTPs and primers were removed using the PCR product purification kit (Marlingen Bioscience, Ijamsville, MD, USA). The cycle sequencing was performed with the Bigdye Terminator v3.1 kit (Applied Biosystems, Foster City, CA, USA). All sequencing reactions were performed according to the manufacturer's instructions using the ABI3130 Genetic Analyzer (Applied Biosystems). The 18S rDNA and rbcl sequences were used in a BLAST search in order to retrieve the most closely related sequences.

Table 1. List of primers used in this study, including the primer sequences, amplicon length, annealing temperature, and the sequencing success rate for four strains tested.

Primer	Molecular Marker	Sequence	Annealing T°	Reference
18SF	18S rDNA	AACCTGGTTGATYCTGCCAG	56 °C, 60 °C	[32]
Euk516r	18S rDNA	ACCAGACTTGCCCTCC	56 °C, 60 °C	[33]
Tetra_rbcL_F	rbcl	GKACTTGGACAACTGTATGGACKGATGGT	56 °C	IFAPA
Tetra_rbcL_R	rbcL	GRTCTTTTTCWACRTAAGCATCACGCATTA	56 °C	IFAPA

2.3. Lipid Analysis

Extraction and Measurement of Lipid Contents

The biomass was harvested at a stationary phase growth (after 15 days of cultivation) by centrifugation (at 4400 rpm for 15 min) and lyophilization for 12 h [30].

The evaluation of lipids fraction from selected microalgae was carried out using the Folch method [35]: 200 mg of lyophilized biomass was extracted in triplicate for 30 min with a chloroform/methanol (2:1, *v/v*) mixture at 25 °C under agitation. The procedure was repeated three times. The organic phase was centrifuged at 3000 RCF for 15 min. The sample was then filtered using Whatman N°1 filter paper in a funnel, collected in a flask, and evaporated at 40 °C.

The productivity of biomass was calculated from the following equation: PB (mg/L/d) = CB/T; where CB (mg/L) was the concentration of biomass from the beginning until the end of the cultivation, and T was the duration of cultivation (15 days).

In addition, the lipid productivity of each sample was determined by Li et al.'s (2008) equation: PL = LT/(V culture × T) and % lipids = (LT/(CB × V culture)) × 100; where PL was the lipid productivity (mg/L/day), LT was the total lipids (mg), T was the duration of the experiment, and V culture was the volume [36,37].

2.4. HPLC Analysis

HPLC-MS analyses of the lipid contents from the obtained residue dissolved in chloroform were performed on a Shimadzu Nexera LC-30A system (Shimadzu, Kyoto, Italy), consisting of a CBM-20A controller, two LC-30AD dual-plunger parallel flow pumps, a DGU-20A5 degasser, a CTO-20A oven, and a SIL-30AC autosampler. The HPLC system was coupled with an LCMS-2020 single quadrupole mass spectrometer equipped with an APCI interface (Shimadzu, Kyoto, Japan). Chromatographic separation was achieved on two serially coupled Ascentis Express C18 columns 100 × 2.1 mm L × I.D., 2.7 µm d.p. (Merck Life Science, Merck KGaA, Darmstadt, Germany), and the injection volume was 10 µL. A linear gradient of ACN/water (80:20, *v/v*) (A) and IPA (B) was run at a mobile phase flow rate of 500 µL/min: 0–105 min, 0–50% B (hold for 5 min). MS parameters were as follows: m/z range, 150–1200; ion accumulation time, 0.6 s; nebulizing gas (N_2) flow rate, during gas flow rate, 2 L/min; detector voltage, 4.5 kV; interface temperature, 450 °C; CDL temperature, 250 °C; block temperature, 300 °C.

2.5. Statistical Analysis

All statistical tests were performed using SPSS statistical software (SPSS software version 16.0 IBM). The different medians of results were analyzed by one-way ANOVA with a significance level of $p < 0.05$ and compared by Tukey's TSD method [38].

3. Results and Discussion

3.1. Isolation of Native Microalgae

The sampling and isolation procedures used in this experiment were successful in establishing a culture collection of ninety-six local microalgae. The microscopic analysis showed that the majority were green algae (Chlorophyta), followed by diatoms and finally cyanobacteria with the lowest number. Sixty-four microalgae were isolated using successive dilution and thirty-two microalgae by inoculation on agar plate (Figure 3A). Following the analysis of variance, ANOVA one-way showed that the difference was insignificant ($p > 0.05$) between the two isolation processes. The two processes used in this experiment turned out to be reproducible to purify microalgal cells. The use of agar plate for isolation was the preferred method for Coccoid and the most soil algae since it represents an axenic culture for direct established without further treatment [28,31].

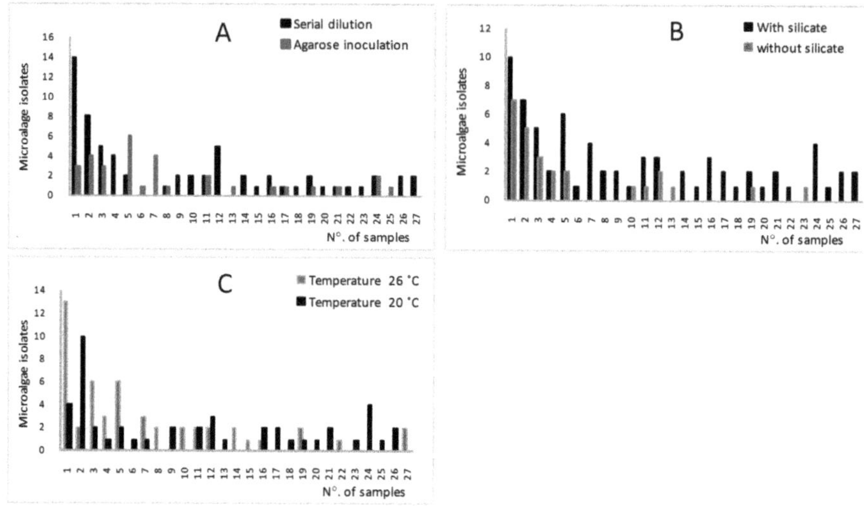

Figure 3. Strains curves for isolated microalgae using successive dilution, (**A**) inoculating on an agar plate, (**B**) in F/2 with or without silicate and (**C**) incubation at 20 °C and 26 °C.

Seventy and twenty-six species were isolated using F/2 culture medium with and without silicate, respectively. The highest number of species were isolated in the culture medium with silicate. The highest number of isolated species was obtained in December 2017 using F/2 medium added by silicate (Figure 3B). In addition, the statistical analysis using One-way ANOVA (Table 2) revealed that the difference between the two groups was highly significant ($p < 0.01$). Therefore, the silicate nutrient added to the F/2 medium had a great impact in regulating the cell growth [39].

Table 2. ANOVA-test statistics values of each isolated microalgae.

One-Factor ANOVA					
Parameter	**Source**	**df**	**Sum of Squares**	**F**	**P**
	Inter-group	1	35.85	9.07	0.004 *
Silicate	Intra-group	52	205.48		
	Total	53	241.33		

* The average difference is significant at the 0.05 level.

Following previous results, the highest number of species, isolated using a culture medium added by silicate, may be explained by the dominance of diatoms. Moreover, Egge and Aksnes (1992) confirmed that diatom strains represent 70% of the cell numbers by using a concentration of 2 mM silicate [40].

Temperature is one of the key criteria for the growth of microalgae and directly acts on the linear and exponential growth of microalgae species [41,42]. In this experiment, fifty species and forty-six strains were isolated at 26 °C and 22 °C, respectively (Figure 3C). Our results were confirmed by Ahlgren's study [41], which showed that the range of temperatures between 16–27 °C was determined as optimal for algal growth rates. Consequently, the species of microalgae were flexible and adaptable with the temperatures tested and able to produce several physiological and biochemical reactions [41–44]. The insignificant difference ($p = 0.823$) between the temperatures confirmed the possibility of growth cells at 26 °C and 22 °C.

3.2. Molecular Identification of Native Microalgae

Four species of microalgae collected from different seawater habitats were identified by microscopic morphological examination based on the form of their cells (Figure 4). Subsequently, two molecular markers were considered to ensure their taxonomic group (Figure 5). The isolated microalgae 1, 4, 5, and 7 were closely related to *Nannochloropsis gaditana* Lubián (MN625926), *Phaeodactylum tricornutum* Bohlin (MN625939), *Nannochloris* sp. KMMCC161 Naumann (MN625923) and *Tetraselmis suecica* Butcher (MN625941) deposited in the NCBI database under GenBank with mentioned accession numbers in parentheses.

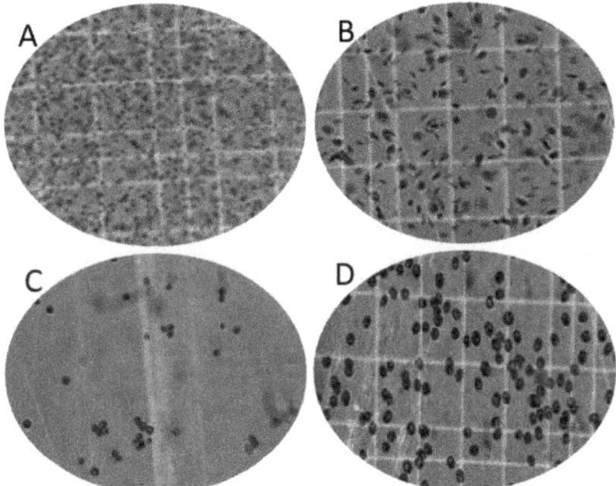

Figure 4. Light microscopic observation of isolated microalgae: (**A**) *Nannochloropsis gaditana*, (**B**) *Phaeodactylum tricornutum*, (**C**) *Nannochloris* sp., and (**D**) *Tetraselmis suecica*.

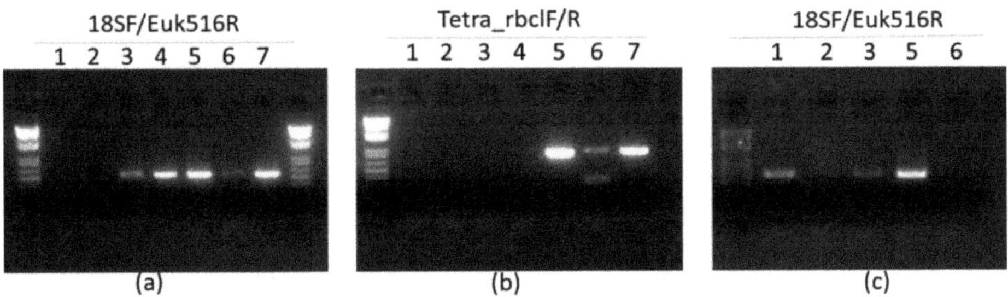

Figure 5. Molecular identification of selected microalgae. 1 to 7: number of strains. (**a**,**b**): annealing T° of 56 °C. (**c**): annealing T° of 60 °C.

3.3. Biomass and Lipid Productivity

One of the key criteria to select microalgae for lipid applications is high intracellular lipid content [45]. Growth and biomass productivity are the most studied parameters to determine the suitable microalgae able to cultivate and using for commercial algal production. In our experiment, a significant difference in lipid levels was determined. The highest biomass productivity was reported in *N. gaditana* then, *Nannochloris* sp., *P. tricornutum*, and *T. suecica* with 97.09, 96.92, 24.47, and 12.54 mg/L/day, respectively. The highest and the lowest levels of lipid productivity were respectively obtained for *Nannochloris* sp., and *T. suecica* species that were measured to be 15.93 and 0.92 mg/L/day (Table 3). Growth rate and biomass productivity directly affect the lipid content [46]; lower biomass productivity can negatively affect productivity even if the lipid content is high [47].

Table 3. Biomass and lipid content of selected microalgae species in batch culture (400 mL) after 15 days of growth in F/2 medium.

Species	Weight of dry biomass (mg/400 mL)	Concentration of biomass (mg/L)	Productivity of biomass (mg/L/day)	Total lipids (mg/400mL)	Lipids (%)	Lipid productivity (mg/L/day)
P. tricornutum	146.82 ± 1.1 [b]	367.05 ± 0.15 [b]	24.47 ± 0.15 [b]	27.85 ± 0.1 [b]	47.43 ± 0.68 [d]	4.64 ± 0.1 [b]
T. suecica	75.25 ± 0.2 [a]	188.13 ± 0.9 [a]	12.54 ± 0.03 [a]	5.50 ± 0.04 [a]	18.28 ± 0.7 [a]	0.92 ± 0.02 [a]
N. gaditana	582.53 ± 0.15 [c]	1456.33 ± 1.43 [c]	97.09 ± 0.68 [c]	33.29 ± 0.8 [c]	14.28 ± 0.32 [b]	5.55 ± 0.01 [b]
Nannochloris sp.,	581.52 ± 1.4 [c]	1453.8 ± 1.3 [c]	96.92 ± 0.72 [c]	95.58 ± 0.81 [d]	41.08 ± 0.26 [c]	15.93 ± 0.9 [c]

Values are medians of three repetitions ± standard deviation. The different letters (a,b,c,d) indicate significant differences between species (Tukey HSD, $p < 0.05$).

The total lipid contents of the microalgae cultured under our experimental growth conditions ranged from 18.28% to 47.43% of their dry weight. *P. tricornutum* and *Nannochloris* sp., showed the highest lipid content at 47.43% and 41.08%, while *T. suecica* and *N. gaditana* have the lowest lipid content at 18.28 and 14.28%, respectively. *Dunaliella tertiolecta* species isolated from Morocco by El Arroussi et al. (2017) had a lipid content of 21%, which confirmed that the lipid productivity was greatly affected by cultivation conditions; salinity, light intensity, temperature, pH of medium, and composition of culture medium used [21,45]. In addition, many studies have confirmed that the limited concentration of nitrogen favored the accumulation of lipids in *Chlorella* species [46–49].

3.4. HPLC Analysis of Lipid Contents

LC-MS analysis was applied to separate and identify the lipid fraction in the four isolated microalgae. The separation of lipids classes reported in Figure 6 confirmed a good separation of neutral and polar lipids. Table 4 lists the details of different lipid classes tentatively identified in all algal samples confirmed by literature data [23,25,50,51]. The identification was carried out with the support of LIPID MAPS Lipidomics Gateway (https://www.lipidmaps.org/), which allowed the identification of a total of 77 compounds in all the selected microalgae species. The lipid composition was constituted basically by mono-, di- and triglycerides (MG, DG, and TG), sulfoquinovosyldiacylglicerols (SQDG), mono- and digalactosyldiacylglycerols (MGDG and DGDG), phospholipids (PLs) and carotenoids (xanthophylls and chlorophylls) in the four microalgal isolates. The lipid fraction of microalgae can also include glycolipids, sterols, carotenoids, chlorophylls, and phospholipids. Algae can synthesize fatty acids to product their membrane lipids in the optimal growth conditions, which constitute phospholipids and glycolipids. However, in conditions of stress, algae alter their biosynthetic pathways and generate neutral lipids to store energy in the form of TGs [52–54]. Notably TGs are significantly produced by the environmental microalgal isolates as demonstrated by the relatively intense peaks of the chromatograms. In fact, intense peaks were identified in *N. gaditana* and *Nannochloris* sp. with TGs C17:0LAr/C17:0OEp—OOP/SLnP and in *T. suecica* with TGs (OOP/PPoG/PoOL), while in *P. tricornutum*, glycolipids SQDG (34:4) occurred as the highest peaks. However, previous studies have demonstrated that some *Tetraselmis* sp., *Nannochloris* sp., *Phaeodactylum tricornutum*, and *Nannochloropsis gaditana* species can produce more lipids under certain stressed conditions [32,47,49,50].

Table 4. Compounds tentatively identified in the lipid fraction.

Peak	Compounds	[M+H]+	[M+H]−	*N. gaditana*	*T. suecica*	*P. tricornutum*	*Nannochloris* sp.,
1	MG 20:0	369.5		+	-	-	-
2	MG 16:0	313.2		+	-	-	-
3	DG (36:4)	599.5		+	+	-	-
4	PG (34:3)		762.3	-	-	+	-
5	DG (34:2)	617.3		+	-	-	-
6	SQDG (32:1)	549.5	791.5	-	-	+	-
7	DG (36:2)	603.4		-	+	-	-
8	DG (32:1)	549.5		-	-	+	-
9	Neoxanthin	601.7		+	-	-	-
10	DG (32:0)	551.4		-	-	-	+

Table 4. *Cont.*

Peak	Compounds	[M+H]⁺	[M+H]⁻	N. gaditana	T. suecica	P. tricornutum	Nannochloris sp.,
11	PG (34:4)		741.4	-	+	-	+
12	PG (34:3)		762.3	-	+	-	-
13	MGDG (34:5)	766.5		+	-	-	-
14	DG (32:5)	583.6		-	+	-	-
15	TG (ArArO/SArEp)	928.5	909.7	+	-	-	-
16	PE (34:3)		762.4	+	-	-	-
17	DG (32:4)	585.5		-	-	-	+
18	DGDG (36:6)	954.6	935.4	+	-	-	-
19	SQDG (34:4)		813.6	-	-	+	-
20	DGDG (34:4)	930.6		-	-	-	+
21	DGDG (36:6)	954.6	935.4	-	-	+	-
22	Antheraxanthin	585.9		-	+	-	-
23	PC (36:1)	788.5	812.7	+	-	+	-
24	SQDG (34:0)		821.5	+	-	-	-
25	PI (36:4)		857.5	+	-	-	-
26	trans-Lutein	569.9		-	-	-	+
27	TG (C20:3LL)	907.6		+	-	-	-
28	SQDG (34:3)		815.5	-	-	+	-
29	TG (20:3LL)	907.5		-	-	+	-
30	PG (34:1)		766.3	-	+	-	-
31	PC (36:3)		784.5	+	+	-	-
32	MGDG (36:5)	794.5		+	-	-	-
33	MGDG (34:6)	764.5		+	-	-	-
34	MGDG (36:6)	792.5		-	-	+	-
35	PG (36:5)		786.5	-	-	-	+
36	PG (34:1)		766.3	-	-	+	-
37	TG (SOAr)	908.6		+	-	-	-
38	PE (39:6)		776.5	+	-	-	-
39	PE (38:3)		768.5	+	-	-	-
40	PG (36:4)		788.4	-	-	-	+
41	DGDG (34:2)	934.5	915.6	-	+	-	+
42	PC (38:3)	814.5		+	-	-	-
43	PG (34:0)		768.5	-	-	+	-
44	SQDG (34:1)		819.4	-	-	+	-
45	PC (38:5)	808.8		-	+	-	-
46	DGDG (36:1)	964.7		-	+	-	-
47	hydroxychlorophyllide b	645.1		-	+	-	-
48	PC (33:2)		744.5	+	-	-	-
49	PE (38:5)		764.5	+	-	-	-
50	b-carotene	537.9		+	-	-	-
51	TG (LnLnPo)	849.4		-	-	-	+
52	MGDG (34:2)	772.5	753.6	-	-	-	+
53	TG (LnGG/OGLn/SGAr)	936.8		-	+	-	-
54	PC (33:1)–PC (O-16:0/18:1)	746.5		-	-	+	-
55	TG (GEpD)	981.5		-	+	-	-
56	PI (40:8)		905.5	-	+	-	-
57	TG (GLL)	909.5		-	+	-	-
58	TG (ArArAr)	950.5		-	+	-	-
59	PC (32:2)		730.5	+	-	-	-
60	TG (OLC18:4)	877.5		-	-	+	-
61	TG (C17:0LAr/C17:0OEp–OOP/SLnP)	893.6–859.8		+	-	-	-
62	TG (EpC18:4C18:4)	893.4		-	+	-	-
63	TG (GLL/LLnA/OLnG)	909.5		-	-	+	-
64	TG (EpEpL)	923.5		-	-	+	-
65	PI (38:1)		892.6	-	-	+	-
66	TG (SSS)	891.4		-	-	+	-
67	Anhydroeschscholtzxanthin		529.3	-	+	-	+
68	TG (C18:4C18:4Ep)	893.6		+	-	-	+
69	TG (OOP/PPoG/PoOL)	925.6		-	+	-	-
70	PC (44:5)		890.3	-	-	+	-
71	TG (EpC18:4C18:4)	893.6		-	-	+	-
72	TG (ArArLn/OEpEp/ArEpL)	925.6		-	+	-	-
73	TG (SOO/SSL/PLA)	887.5		+	-	-	-
74	TG (MMDh)	823.5		-	+	-	-
75	TG (SOO/SSL/PLA)	887.5		-	-	+	-
76	TG (OOMo/LnLn18:4)	871.4		+	-	-	+
77	Pheophytin a (C₅₅H₇₄N₄₀₅)	871.4	870.6	+	-	+	+

Note: MG, monoglyceride; DG, diglyceride; TG, triglyceride; SQDG, sulfoquinovosyldiacylglicerol; MGDG, monogalactosyldiacylglycerol; DGDG, digalactosyldiacylglycerol; PG, phosphatidylglycerol; PE, phosphatidylethanolamine; PC, phosphatidylcholine; PI, phosphatidylinositol. M = C14:0; Mo = C14:1; P = C16:0; Po = C16:1; S = C18:0; O = C 18:1; L = C18:2; Ln = C18:3; A = C20:0;0020G = C20:1; Ar = C20:4; Ep = C20:5; B = C22:0; D = C22:5; Dh = C22:6.

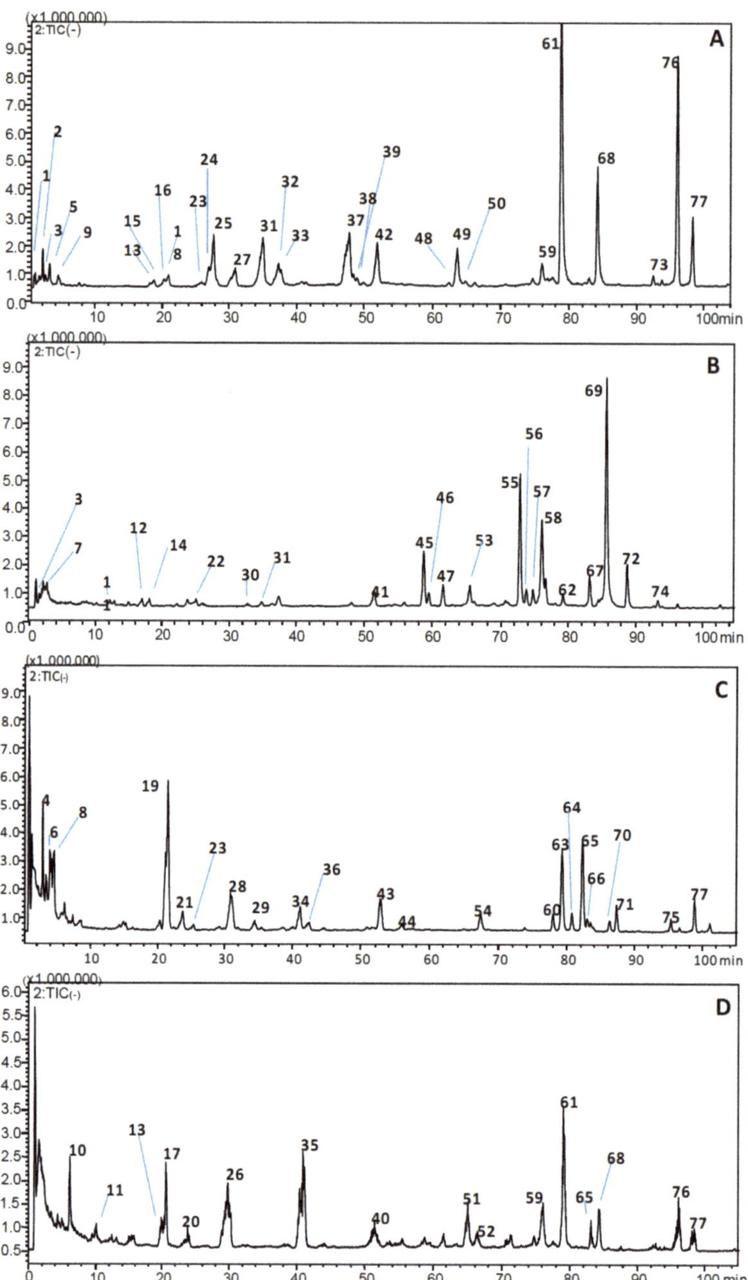

Figure 6. Base peak chromatograms of lipids extract of four isolated microalgae. (**A**) *N. gaditana*; (**B**); *T. suecica*; (**C**) *P. tricornutum*; (**D**) *Nannochloris* sp. For experimental conditions, see the text. For peak identification, see Table 4.

The comprehensive lipid profiles obtained for the four strains showed their higher potential of glycolipids, phospholipids, and triacylglycerols to be an ideal source of food additives, ingredient for functional foods, and nutraceuticals applications [31,51].

4. Conclusions

This study reports the identification of four strains of microalgae from Moroccan Mediterranean seawater, which might be suitable for lipid production to be potentially used as supplement aquatic and animal feed rich in PUFA and other food products with higher omega-3 fatty acids content. In particular, the data from this study showed that *Nannochloris* sp., had the highest lipid productivity of 15.93 mg/L/day. Furthermore, the HPLC-MS analysis showed their highest content of lipidic molecules (77 in total) and might be useful as dietary supplements or biofuels feedstock.

Author Contributions: Conceptualization, F.C. and M.M.; methodology, K.C.; software, H.C. and A.E.; validation, F.R. and M.M.; formal analysis, D.M. and M.O., F.R.; investigation, M.H.; writing—original draft preparation, I.H.; writing—review and editing, F.C.; supervision, N.S.S., L.M., F.C., and J.A. All authors have read and agreed to the published version of the manuscript.

Funding: This research received no external funding.

Acknowledgments: The authors are thankful to Shimadzu and MerckLife Science Corporations for their continuous support.

Conflicts of Interest: No potential conflict of interest was reported by the authors.

References

1. Richmond, A. *Handbook of Microalgal Culture: Biotechnology and Applied Phycology*; Blackwell Science: Oxford, UK, 2004; p. 289; ISBN 978-1-4051-2832-2.
2. Falkowski, P.G. The Evolution of Modern Eukaryotic Phytoplankton. *Science* **2004**, *305*, 354–360. [CrossRef] [PubMed]
3. Sathasivam, R.; Radhakrishnan, R.; Hashem, A.; Abd Allah, E.F. Microalgae metabolites: A rich source for food and medicine. *Saudi J. Biol. Sci.* **2017**, *26*, 709–722. [CrossRef] [PubMed]
4. Craggs, R.J.; Adey, W.H.; Jenson, K.R.; John, M.S.S.; Green, F.B.; Oswald, W.J. Phosphorus removal from wastewater using an algal turf scrubber. *Water Sci. Technol.* **1996**, *33*, 191–198. [CrossRef]
5. Demirbas, A.; Demirbas, M.F. Importance of algae oil as a source of biodiesel. *Energy Convers. Manag.* **2011**, *52*, 163–170. [CrossRef]
6. Enzing, C.; Ploeg, M.; Barbosa, M.; Sijtsma, L. Microalgae-based products for the food and feed sector: An outlook for Europe. *JRC Sci. Policy Rep.* **2014**, 19–37.
7. Spolaore, P.; Joannis-Cassan, C.; Duran, E.; Isambert, A. Commercial applications of microalgae. *J. Biosci. Bioeng.* **2006**, *101*, 87–96. [CrossRef]
8. Hu, G.-P.; Yuan, J.; Sun, L.; She, Z.-G.; Wu, J.-H.; Lan, X.-J.; Zhu, X.; Lin, Y.-C.; Chen, S.-P. Statistical research on marine natural products based on data obtained between 1985 and 2008. *Mar. Drugs.* **2011**, *9*, 514–525. [CrossRef]
9. De JesusRaposo, M.F.; De Morais, R.M.S.C.; de Morais, B.; Miranda, A.M. Bioactivity and applications of sulphated polysaccharides from marine microalgae. *Mar. Drugs.* **2013**, *11*, 233–252.
10. Peng, J.; Yuan, J.-P.; Wu, C.-F.; Wang, J.-H. Fucoxanthin, a marine carotenoid present in brown seaweeds and diatoms: Metabolism and bioactivities relevant to human health. *Mar. Drugs.* **2011**, *9*, 1806–1828. [CrossRef] [PubMed]
11. Haimeur, A.; Ulmann, L.; Mimouni, V.; Guéno, F.; Pineau-Vincent, F.; Meskini, N.; Tremblin, G. The role of Odontellaaurita, a marine diatom rich in EPA, as a dietary supplement in dyslipidemia, platelet function and oxidative stress in high-fat fed rats. *Lipids Health Dis.* **2012**, *11*, 147. [CrossRef]
12. Goiris, K.; Muylaert, K.; Fraeye, I.; Foubert, I.; De Brabanter, J.; De Cooman, L. Antioxidant potential of microalgae in relation to their phenolic and carotenoid content. *J. Appl. Phycol.* **2012**, *24*, 1477–1486. [CrossRef]
13. Nuzzo, G.; Gallo, C.; d'Ippolito, G.; Cutignano, A.; Sardo, A.; Fontana, A. Composition and quantitation of microalgal lipids by ERETIC 1H NMR method. *Mar. Drugs.* **2013**, *11*, 3742–3753. [CrossRef]
14. Hannon, M.; Gimpel, J.; Tran, M.; Rasala, B.; Mayfield, S. Biofuels from algae: Challenges and potential. *Biofuels* **2010**, *1*, 763–784. [CrossRef]
15. Elliott, L.G.; Feehan, C.; Laurens, L.M.; Pienkos, P.T.; Darzins, A.; Posewitz, M.C. Establishment of a bioenergy-focused microalgal culture collection. *Algal Res.* **2012**, *1*, 102–113. [CrossRef]

16. Abdelaziz, A.E.M.; Leite, G.B.; Belhaj, M.A.; Hallenbeck, P.C. Screening microalgae native to Quebec for wastewater treatment and biodiesel production. *Bioresour. Technol.* **2014**, *157*, 140–148. [CrossRef]

17. Poblete-Castro, I.; Escapa, I.F.; Jäger, C.; Puchalka, J.; Lam, C.M.C.; Schomburg, D.; Prieto, M.A.; dos Santos, V.A.M. The metabolic response of P. putida KT2442 producing high levels of polyhydroxyalkanoate under single-and multiple-nutrient-limited growth: Highlights from a multi-level omics approach. *Microb. Cell Factor.* **2012**, *11*, 34. [CrossRef] [PubMed]

18. Beetul, K.; Gopeechund, A.; Kaullysing, D.; Mattan-Moorgawa, S.; Puchooa, D.; Bhagooli, R. *Challenges and Opportunities in the Present Era of Marine Algal Applications*; InTechOpen: London, UK, 2016; pp. 237–276.

19. Judd, S.; van den Broeke, L.J.P.; Shurair, M.; Kuti, Y.; Znad, H. Algal remediation of CO_2 and nutrient discharges: A review. *Water Res.* **2015**, *87*, 356–366. [CrossRef] [PubMed]

20. Ismail, N.M.; Ismail, A.F.; Mustafa, A.; Matsuura, T.; Soga, T.; Nagata, K.; Asaka, T. Qualitative and quantitative analysis of intercalated and exfoliated silicate layers in asymmetric polyethersulfone/cloisite15A® mixed matrix membrane for CO_2/CH_4 separation. *Chem. Eng. J.* **2015**, *268*, 371–383. [CrossRef]

21. Guschina, I.A.; Harwood, J.L. Lipids and lipid metabolism in eukaryotic algae. *Prog. Lipid Res.* **2006**, *45*, 160–186. [CrossRef] [PubMed]

22. Ragonese, C.; Tedone, L.; Beccaria, M.; Torre, G.; Cichello, F.; Cacciola, F.; Dugo, P.; Mondello, L. Characterisation of lipid fraction of marine macroalgae by means of chromatography techniques coupled to mass spectrometry. *Food Chem.* **2014**, *145*, 932–940. [CrossRef]

23. He, H.; Rodgers, R.P.; Marshall, A.G.; Hsu, C.S. Algae polar lipids characterized by online liquid chromatography coupled with hybrid linear quadrupole ion trap/fourier transform ion cyclotron resonance mass spectrometry. *Energy Fuels.* **2011**, *25*, 4770–4775. [CrossRef]

24. Rigano, F.; Oteri, M.; Russo, M.; Dugo, P.; Mondello, L. Proposal of a Linear Retention Index System for Improving Identification Reliability of Triacylglycerol Profiles in Lipid Samples by Liquid Chromatography Methods. *Anal. Chem.* **2018**, *90*, 3313–3320. [CrossRef] [PubMed]

25. Beccaria, M.; Inferrera, V.; Rigano, F.; Gorynski, K.; Purcaro, G.; Pawliszyn, J.; Dugo, P.; Mondello, L. Highly informative multiclass profiling of lipids by ultra-high performance liquid chromatography—Low resolution (quadrupole) mass spectrometry by using electrospray ionization and atmospheric pressure chemical ionization interfaces. *J. Chromatogr. A* **2017**, *1509*, 69–82. [CrossRef] [PubMed]

26. Morowvat, M.H.; Ghasemi, Y. Cell Growth, Lipid Production and Productivity in Photosynthetic Micro-alga Chlorella vulgaris under Different Nitrogen Concentrations and Culture Media Replacement. *RecentPat. Food Nutr. Agric.* **2018**, *9*, 142–151.

27. Brahamsha, B. A genetic manipulation system for oceanic cyanobacteria of the genus *Synechococcus*. *Appl. Environ. Microbiol.* **1996**, *62*, 1747–1751. [CrossRef]

28. Singh, P.; Gupta, S.K.; Guldhe, A.; Rawat, I.; Bux, F. Microalgae Isolation and Basic Culturing Techniques. In *Handbook of Marine Microalgae*; Elsevier: Amsterdam, The Netherlands, 2015; pp. 43–54; ISBN 978-0-12-800776-1.

29. Andersen, R.A. (Ed.) *Algal Culturing Techniques*; Academic Press: Cambridge, UK, 2005; ISBN 978-0-12-088426-1.

30. Tischer, R.G. Pure Culture of Anabaena flos-aquae A-37. *Nature* **1965**, *205*, 419–420. [CrossRef]

31. Izadpanah, M.; Gheshlaghi, R.; Mahdavi, M.A.; Elkamel, A. Effect of light spectrum on isolation of microalgae from urban wastewater and growth characteristics of subsequent cultivation of the isolated species. *Algal Res.* **2018**, *29*, 154–158. [CrossRef]

32. Sakata, T.; Yoshikawa, T.; Maeda, K.; del Castillo, C.S.; Dureza, L.A. Identification of microalgae isolated from green water of tilapia culture ponds in the Philippines. *Mem. Fac. Fish. Kagoshima Univ.* **2005**, *54*, 35–44.

33. Díez, B.; Pedrós-Alió, C.; Marsh, T.L.; Massana, R. Application of Denaturing Gradient Gel Electrophoresis (DGGE) To Study the Diversity of Marine Picoeukaryotic Assemblages and Comparison of DGGE with Other Molecular Techniques. *Appl. Environ. Microbiol.* **2001**, *67*, 2942–2951. [CrossRef]

34. Rodríguez-Ferri, E.; Badiola-Díez, J.J.; Cepeda-Sáez, A.; Domínguez-Rodríguez, L.; Otero-Carballeira, A.; Zurera-Cosano, G. Grupo de trabajo. Informe del ComitéCientífico de la Agencia Española de SeguridadAlimentaria y Nutrición (AESAN) sobre la evisceración de loslagomorfos. *RevistaComitéCientífico AESAN* **2009**, *9*, 31–38.

35. Folch, J.; Lees, M.; Stanley, G.S. A simple method for the isolation and purification of total lipides from animal tissues. *J. Biol. Chem.* **1957**, *226*, 497–509.

36. Bligh, E.G.; Dyer, W.J. A rapid method of total lipid extraction and purification. *Can. J. Biochem. Physiol.* **1959**, *37*, 911–917. [CrossRef]

37. Li, Y.; Horsman, M.; Wang, B.; Wu, N.; Lan, C.Q. Effects of nitrogen sources on cell growth and lipid accumulation of green alga *Neochlorisoleoabundans. Appl. Microbiol. Biotechnol.* **2008**, *81*, 629–636. [CrossRef]

38. Montgomery, D.C. *Design and Analysis of Experiments*; John Wiley & Sons: Hoboken, NJ, USA, 2017; ISBN 1-119-11347-4.

39. Egge, J.K. Are diatoms poor competitors at low phosphate concentrations? *J. Mar. Syst.* **1998**, *16*, 191–198. [CrossRef]

40. Egge, J.K.; Aksnes, D.L. Silicate as regulating nutrient in phytoplankton competition. *Mar. Ecol. Prog. Ser. Oldendorf.* **1992**, *83*, 281–289. [CrossRef]

41. Ahlgren, G. Temperature Functions in Biology and Their Application to Algal Growth Constants. *Oikos* **1987**, *49*, 177. [CrossRef]

42. Thompson, P.A.; Guo, M.; Harrison, P.J.; Whyte, J.N.C. Effects of variation in temperature. ii. on the fatty acid composition of eight species of marine phytoplankton. *J. Phycol.* **1992**, *28*, 488–497. [CrossRef]

43. Stirk, W.A.; Bálint, P.; Tarkowská, D.; Strnad, M.; van Staden, J.; Ördög, V. Endogenous brassinosteroids in microalgae exposed to salt and low temperature stress. *Eur. J. Phycol.* **2018**, *53*, 273–279. [CrossRef]

44. Kang, E.J.; Kim, K.Y. Effects of future climate conditions on photosynthesis and biochemical component of Ulva pertusa (Chlorophyta). *Algae* **2016**, *31*, 49–59. [CrossRef]

45. El Arroussi, H.; Benhima, R.; El Mernissi, N.; Bouhfid, R.; Tilsaghani, C.; Bennis, I.; Wahby, I. Screening of marine microalgae strains from Moroccan coasts for biodiesel production. *Renew. Energy.* **2017**, *113*, 1515–1522. [CrossRef]

46. Griffiths, M.J.; Harrison, S.T. Lipid productivity as a key characteristic for choosing algal species for biodiesel production. *J. Appl. Phycol.* **2009**, *21*, 493–507. [CrossRef]

47. Rodolfi, L.; ChiniZittelli, G.; Bassi, N.; Padovani, G.; Biondi, N.; Bonini, G.; Tredici, M.R. Microalgae for oil: Strain selection, induction of lipid synthesis and outdoor mass cultivation in a low-cost photobioreactor. *Biotechnol. Bioeng.* **2009**, *102*, 100–112. [CrossRef]

48. Haixing, C.; Qian, F.; Yun, H.; Ao, X.; Qiang, L.; Xun, Z. Improvement of microalgae lipid productivity and quality in an ion-exchange-membrane photobioreactor using real municipal wastewater. *Int. J. Agric. Biol. Eng.* **2017**, *10*, 97–106.

49. Illman, A.M.; Scragg, A.H.; Shales, S.W. Increase in Chlorella strains calorific values when grown in low nitrogen medium. *Enzyme Microb. Technol.* **2000**, *27*, 631–635. [CrossRef]

50. MacDougall, K.M.; McNichol, J.; McGinn, P.J.; O'Leary, S.J.; Melanson, J.E. Triacylglycerol profiling of microalgae strains for biofuel feedstock by liquid chromatography–high-resolution mass spectrometry. *Anal. Bioanal. Chem.* **2011**, *401*, 2609–2616. [CrossRef]

51. Alves, E.; Domingues, M.R.M.; Domingues, P. Polar lipids from olives and olive oil: A review on their identification, significance and potential biotechnological applications. *Foods* **2018**, *7*, 109. [CrossRef]

52. Hu, Q.; Sommerfeld, M.; Jarvis, E.; Ghirardi, M.; Posewitz, M.; Seibert, M.; Darzins, A. Microalgaltriacylglycerols as feedstocks for biofuel production: Perspectives and advances. *Plant J.* **2008**, *54*, 621–639. [CrossRef] [PubMed]

53. Wang, B.; Li, Y.; Wu, N.; Lan, C.Q. CO_2biomitigation using microalgae. *Appl. Microbiol. Biotechnol.* **2008**, *79*, 707–718. [CrossRef]

54. Chisti, Y. Biodiesel from microalgae. *Biotechnol. Adv.* **2007**, *25*, 294–306. [CrossRef]

Publisher's Note: MDPI stays neutral with regard to jurisdictional claims in published maps and institutional affiliations.

 *foods*

Article

Simultaneous Determination of 6-Shogaol and 6-Gingerol in Various Ginger (*Zingiber officinale* Roscoe) Extracts and Commercial Formulations Using a Green RP-HPTLC-Densitometry Method

Ahmed I. Foudah [1], Faiyaz Shakeel [2], Hasan S. Yusufoglu [1], Samir A. Ross [3,4] and Prawez Alam [1,*]

1   Department of Pharmacognosy, College of Pharmacy, Prince Sattam Bin Abdulaziz University, Al-Kharj 11942, Saudi Arabia; a.foudah@psau.edu.sa (A.I.F.); h.yusufoglu@psau.edu.sa (H.S.Y.)
2   Department of Pharmaceutics, College of Pharmacy, King Saud University, Riyadh 11451, Saudi Arabia; faiyazs@fastmail.fm
3   National Center for Natural Products Research, University of Mississippi, Oxford, MS 38677, USA; sroos@olemiss.edu
4   Department of Biomolecular Sciences, School of Pharmacy, University of Mississippi, Oxford, MS 38677, USA
*   Correspondence: p.alam@psau.edu.sa

Received: 14 July 2020; Accepted: 11 August 2020; Published: 18 August 2020

Abstract: Various analytical methodologies have been reported for the determination of 6-shogaol (6-SHO) and 6-gingerol (6-GIN) in ginger extracts and commercial formulations. However, green analytical methods for the determination of 6-SHO and 6-GIN, either alone or in combination, have not yet been reported in literature. Hence, the present study was aimed to develop a rapid, simple, and cheaper green reversed phase high-performance thin-layer chromatography (RP-HPTLC) densitometry method for the simultaneous determination of 6-SHO and 6-GIN in the traditional and ultrasonication-assisted extracts of ginger rhizome, commercial ginger powder, commercial capsules, and commercial ginger teas. The simultaneous analysis of 6-SHO and 6-GIN was carried out via RP-18 silica gel 60 F254S HPTLC plates. The mixture of green solvents, i.e., ethanol:water (6.5:3.5 *v/v*) was utilized as a mobile phase for the simultaneous analysis of 6-SHO and 6-GIN. The analysis of 6-SHO and 6-GIN was performed at λ_{max} = 200 nm for 6-SHO and 6-GIN. The densitograms of 6-SHO and 6-GIN from traditional and ultrasonication-assisted extracts of ginger rhizome, commercial ginger powder, commercial capsules, and commercial ginger teas were verified by obtaining their single band at R_f = 0.36 ± 0.01 for 6-SHO and R_f = 0.53 ± 0.01 for 6-GIN, compared to standard 6-SHO and 6-GIN. The green RP-HPTLC method was found to be linear, in the range of 100–700 ng/band with R^2 = 0.9988 for 6-SHO and 50–600 ng/band with R^2 = 0.9995 for 6-GIN. In addition, the method was recorded as "accurate, precise, robust and sensitive" for the simultaneous quantification of 6-SHO and 6-GIN in traditional and ultrasonication-assisted extracts of ginger rhizome, commercial ginger powder, commercial capsules, and commercial ginger teas. The amount of 6-SHO in traditional extracts of ginger rhizome, commercial ginger powder, commercial capsules, and commercial ginger teas was obtained as 12.1, 17.9, 10.5, and 9.6 mg/g of extract, respectively. However, the amount of 6-SHO in ultrasonication-assisted extracts of ginger rhizome, commercial ginger powder, commercial capsules, and commercial ginger teas were obtained as 14.6, 19.7, 11.6, and 10.7 mg/g of extract, respectively. The amount of 6-GIN in traditional extracts of ginger rhizome, commercial ginger powder, commercial capsules, and commercial ginger teas were found as 10.2, 15.1, 7.3, and 6.9 mg/g of extract, respectively. However, the amount of 6-GIN in ultrasonication-assisted extracts of ginger rhizome, commercial ginger powder, commercial capsules, and commercial ginger teas were obtained as 12.7, 17.8, 8.8, and 7.9 mg/g of extract, respectively. Overall, the results of this study indicated that the proposed analytical technique could be effectively used for the simultaneous quantification of 6-SHO and 6-GIN in a wide range of plant extracts and commercial formulations.

Keywords: 6-gingerol; 6-shogaol; commercial formulation; ginger extract; green reversed phase high-performance thin-layer chromatography (RP-HPTLC); simultaneous analysis

---

1. Introduction

The roots or rhizomes of ginger (*Zingiber officinale* Roscoe; family: Zingeberaceae) have been used as dietary supplements since ancient times [1]. Although it is cultivated in several countries around the globe, India and China are the leading producers of ginger [1,2]. In recent years, ginger has gained attention from researchers around the globe due to its broad range of therapeutic activities, in addition to its low toxicity [1–3]. The main therapeutic activities of ginger are antioxidant [3], anti-inflammatory [4], anti-apoptotic [5], analgesic [4], cytotoxic [5], anti-proliferative [5,6], antitumor [6], and anti-platelets [7] activities. The therapeutic activities of ginger are due to the presence of biomarker compounds, such as various gingerols and shogaols [6,7]. The most abundant gingerols of ginger rhizome are 6-gingerol (6-GIN), 8-GIN, and 10-GIN [1,7]. Among shogaols, the most abundant is 6-shogaol (6-SHO) [6]. The chemical structures/formulae of 6-SHO and 6-GIN are presented in Figure 1.

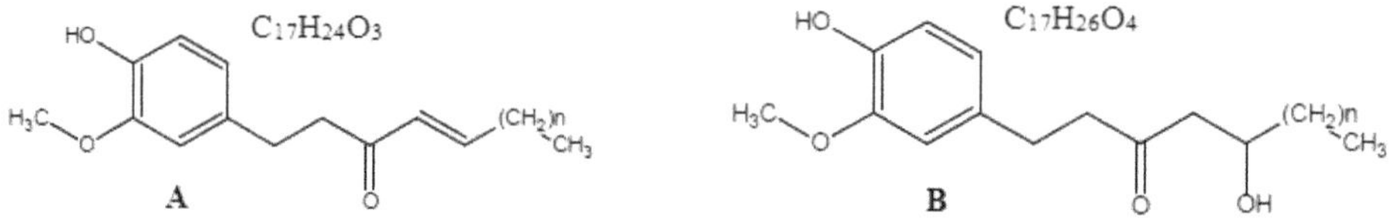

Figure 1. Chemical structure of (**A**) 6-shogaol (6-SHO) and (**B**) 6-gingerol (6-GIN).

A single ultra-violet (UV) spectrometry method was reported for the determination of 6-GIN in ginger extract [8]. Various high-performance liquid chromatography (HPLC) methods were reported for the determination of 6-GIN in plant extracts, commercial food products, and commercial formulations [1,2,9–14]. Different liquid chromatography mass-spectrometry (LC-MS) methods were applied for the analysis of 6-GIN, either alone or in combination with other ginger compounds in plant extracts and commercial formulations [15–19]. Some high-performance thin layer chromatography (HPTLC) methods have also been reported for the determination of 6-GIN in ginger extract, commercial foods, and commercial formulations [20–22]. The HPTLC method had also been utilized for the simultaneous determination of 6-GIN, 8-GIN, and 10-GIN in ultrasonication-assisted extract of ginger [23]. The HPTLC technique was also applied for the determination of a similar compound 8-GIN in plant extracts, commercial foods, and commercial formulations [24]. Some HPLC methods have been applied for the simultaneous determination of 6-SHO and 6-GIN in ginger extract and commercial formulations [25–28]. A HPLC method was also proposed for the simultaneous determination of 6-SHO and 6-GIN in human plasma samples [29]. A LC-MS method was applied for the simultaneous determination of 6-SHO and 6-GIN in ginger extract [30]. Although various analytical methodologies have been reported for the determination of 6-GIN or the simultaneous determination of 6-SHO and 6-GIN in ginger extracts, commercial foods, and commercial formulations, none of them used green analytical methodology. Moreover, reported HPTLC methods have also used toxic solvents in their mobile phase [20–23]. In addition, most of the reported HPTLC methods are based on normal phase chromatography [20,24]. Few green HPTLC methods, utilizing the same mobile phase (ethanol and water), have been reported for the determination of rivaroxaban, delafloxacin, and diosmin, but the method application and analyzed compounds are different, with respect to the present method [31–33]. Some HPTLC methods have also been reported for the determination of bioactive compounds, such as vanillin and flavonoids in plant extracts, but they used toxic mobile phase in addition to different analyzed compounds [34,35]. Recently, the analytical methods associated with green analytical chemistry or environmentally benign analytical chemistry have increased significantly in literature [31,36,37].

Based on reported literature, it was found that green HPTLC methods have not been reported for the analysis of 6-SHO or 6-GIN alone or in their combination. Moreover, green reversed phase HPTLC (RP-HPTLC) methods offer many advantages over conventional HPTLC methods, such as the avoidance of nonpolar fractions, avoidance of interference of the impurities, formation of compact spots, detection clarity, and non-toxicity to the environment [36–38]. Hence, the aim of this study was to develop and validate a green RP-HPTLC method for the simultaneous determination of 6-SHO and 6-GIN in traditional and ultrasonication-assisted extracts of ginger rhizome, commercial ginger powder, commercial capsules, and commercial ginger teas for the first time. The proposed green RP-HPTLC method for the simultaneous determination of 6-SHO and 6-GIN was validated for linearity, accuracy, precision, robustness, sensitivity, and specificity, as per International Conference on Harmonization (ICH) Q2 (R1) recommendations [39].

2. Materials and Methods

2.1. Materials

Standard 6-SHO, 6-GIN, and commercial ginger powder were obtained from Natural Remedies (Bangalore, India). HPLC grades of methanol and ethanol were acquired from E-Merck (Darmstadt, Germany). Ginger commercial capsules and ginger roots containing teas were obtained from the local market in Riyadh, Saudi Arabia. Ginger rhizomes were purchased from the local market in Al-Kharj, Saudi Arabia. HPLC grade water was collected from the Milli-Q unit. All the solvents were of chromatography grades and other reagents were of analytical reagent grades.

2.2. Instrumentation and Analytical Conditions

Simultaneous RP-HPTLC-densitometry analysis of 6-SHO and 6-GIN was carried out using the following instrumentation and analytical conditions:

HPTLC apparatus: CAMAG TLC system (CAMAG, Muttenz, Switzerland)

Software: WinCAT's (version 1.4.3.6336, CAMAG, Muttenz, Switzerland)

Syringe for sample application: CAMAG microliter Syringe (Hamilton, Bonaduz, Switzerland)

TLC plate: 10 × 20 cm glass backed plates pre-coated with RP silica gel 60 F254S plates (E-Merck, Darmstadt, Germany)

Sample applicator: CAMAG Linomat-V (CAMAG, Muttenz, Switzerland)

Gas for sample application: nitrogen

Development chamber: CAMAG automatic developing chamber 2 (ADC2) (CAMAG, Muttenz, Switzerland)

TLC scanner: CAMAG TLC scanner-III (CAMAG, Muttenz, Switzerland)

Stationary phase: 10 × 20 cm glass backed plates pre-coated with RP silica gel 60 F254S plates (E-Merck, Darmstadt, Germany)

Mobile phase for 6-SHO and 6-GIN: ethanol:water (6.5:3.5 *v/v*)

Saturation time: 30 min at 22 °C

Development distance on plate: 80 mm

Development mode: linear ascending mode

Sample application rate: 150 nL/s

Densitometry of scanning mode: absorbance/reflectance.

Scanning wavelength for 6-SHO and 6-GIN: 200 nm

2.3. Calibration Curve of 6-SHO and 6-GIN

The stock solution (SS) of 6-SHO and 6-GIN was prepared separately by dissolving an accurately weighed 10 mg of 6-SHO and 6-GIN in 10 mL of methanol. About 1 mL of SS of 6-SHO and 6-GIN was diluted further using mobile phase to obtain the final SS of 100 µg/mL. Serial dilutions of SS of 6-SHO and 6-GIN were made by taking different volumes of 6-SHO SS or 6-GIN SS and diluting them with mobile phase to obtain concentrations in the range of 100–700 ng/band for 6-SHO and 50–600 ng/band for 6-GIN. The SSs of 6-SHO and 6-GIN were prepared in six replicates (n = 6). Around 200 µL of each concentration of 6-SHO and 6-GIN was applied on TLC plates, and the peak area was recorded. The calibration curve (CC) of 6-SHO and 6-GIN was obtained by plotting the concentrations against the measured areas of 6-SHO and 6-GIN. The CCs for 6-SHO and 6-GIN were obtained for six replicates (n = 6).

2.4. Extraction Procedure for Ginger Rhizomes

Accurately weighed 5 g of the dried whole rhizomes of ginger (*Zingiber officinale* Roscoe) were refluxed with methanol (100 mL) for 1 h in a water bath and filtered through Whatman filter paper (No. 41). The marc left out was refluxed again three times with 70 mL of methanol for 1 h and filtered. The solvent was evaporated using a rotary vacuum evaporator, and the residue was dissolved in 50 mL methanol in a volumetric flask. This procedure was performed in three replicates (n = 3). This solution was used as a test solution in the RP-HPTLC-densitometry analysis of 6-SHO and 6-GIN in the methanolic extract of ginger rhizome.

2.5. Extraction Procedure from Commercial Ginger Powder, Capsules, and Ginger Teas

For the simultaneous determination of 6-SHO and 6-GIN in commercial dietary supplement capsules, five soft gelatin capsules containing ginger powder were opened, transferred in a beaker, and mixed to ensure that a homogenous sample was obtained. The ginger root teas and ginger powder were also transferred in a beaker separately. About 1 g of each of ginger root dietary supplement capsule, ginger powder, and tea were weighed and transferred to separate beakers. They were then extracted thrice with 70 mL of methanol separately. Filtrates were combined and concentrated using a rotary vacuum evaporator to a final volume of 10 mL. Each procedure was performed in three replicates (n = 3). The obtained solutions were used as test solutions in the RP-HPTLC analysis. The amount of 6-SHO and 6-GIN was determined in both of the test solutions, using the RP-HPTLC-densitometry method.

2.6. Ultrasonic Extraction Procedure for Whole Rhizomes of Ginger

The ultrasonic extraction of the dried whole rhizomes of ginger was carried out by ultrasonic vibrations (ultrasound-assisted extraction) using the Bransonic series (Model CPX5800H-E; New Jersey, NJ, USA). A total of 5 g of dried whole rhizomes of ginger was taken and extracted with 100 mL of methanol. The solvent was evaporated using a rotary vacuum evaporator and the residue was dissolved in 50 mL methanol in a volumetric flask. It was ultrasonicated at 50 °C for 1 h. This procedure was performed in three replicates (n = 3). This solution was used as a test solution for the simultaneous determination of 6-SHO and 6-GIN in an ultrasound-assisted extract of ginger rhizomes.

2.7. Ultrasonic Extraction Procedure from Commercial Ginger Powder, Capsules, and Ginger Teas

For the simultaneous determination of 6-SHO and 6-GIN in commercial dietary supplement capsules, five capsules containing ginger powder were opened, transferred in a beaker, and mixed to ensure that a homogenous sample was obtained. The ginger teas and ginger powder were also transferred in separate beakers. About 1 g each of ginger root dietary supplement capsule, ginger powder, and tea were weighed and transferred to separate beakers. It was then ultrasonicated thrice at 50 °C for 1 h with 70 mL of methanol separately. Filtrates were combined and concentrated using a rotary vacuum evaporator to a final volume of 10 mL. Each procedure was performed in three replicates (n = 3). The obtained solutions were used as test solutions in the RP-HPTLC analysis. The amount of 6-SHO and 6-GIN in ultrasonic-assisted extracts of commercial capsules and ginger root teas was determined using the RP-HPTLC-densitometry method.

2.8. Method Validation

The proposed green RP-HPTLC-densitometry method for the simultaneous determination of 6-SHO and 6-GIN was validated for linearity, precision, accuracy, robustness, sensitivity, and specificity, according to ICH Q2 (R1) recommendations [39]. The linearity of 6-SHO and 6-GIN was obtained by plotting the concentration of 6-SHO and 6-GIN against their respective HPTLC responses. The linearity for 6-SHO was studied in the range of 100–700 ng/band in six replicates (n = 6). However, the linearity for 6-GIN was recorded in the range of 50–600 ng/band in six replicates (n = 6). The method accuracy for 6-SHO and 6-GIN was determined as the percent of recovery (% recovery) using the standard addition method. The previously analyzed solutions of 6-SHO (100 ng/band) and 6-GIN (100 ng/band) were spiked with extra 0, 50, 100, and 150% amounts of 6-SHO and 6-GIN. The resultant concentrations of 100, 150, 200, and 250 ng/band for 6-SHO and 6-GIN were re-analyzed using a green RP-HPTLC-densitometry method in six replicates (n = 6). The % recovery was calculated at each concentration of 6-SHO and 6-GIN.

Method precision for 6-SHO and 6-GIN was determined as repeatability and intermediate precision. Repeatability (intra-day precision) for 6-SHO and 6-GIN was evaluated by the analysis of samples on the same day at 100, 150, 200, and 250 ng/band concentrations in six replicates (n = 6). However, intermediate (inter-day precision) for 6-SHO and 6-GIN was evaluated using the analysis of samples on three different days at 100, 150, 200, and 250 ng/band concentrations in six replicates (n = 6).

Method robustness for 6-SHO and 6-GIN was determined by introducing some minor modifications in the composition of green mobile phase during 6-SHO and 6-GIN analysis. The original mobile phase composition of ethanol:water (6.5:3.5) was changed into ethanol:water (6.7:3.3) and ethanol:water (6.3:3.7) for the positive and negative level, respectively. The robustness for 6-SHO and 6-GIN was evaluated in six replicates (n = 6).

Method sensitivity for 6-SHO and 6-GIN was determined as the limit of detection (LOD) and limit of quantification (LOQ) by adopting the standard deviation (SD) method. The LOD and LOQ values for 6-SHO and 6-GIN were calculated in six replicates (n = 6) using the following equations:

$$\mathrm{LOD} = 3.3 \times \frac{\mathrm{SD}}{\mathrm{S}} \tag{1}$$

$$\mathrm{LOQ} = 10 \times \frac{\mathrm{SD}}{\mathrm{S}} \tag{2}$$

where S is the slope of the CC of 6-SHO or 6-GIN.

Method specificity for 6-SHO and 6-GIN was evaluated by comparing the retardation factor (R_f) values and UV-absorption spectra of 6-SHO and 6-GIN in the ginger rhizome extract, commercial capsules, and ginger root teas with that of standard 6-SHO and 6-GIN.

2.9. Application of a Green RP-HPTLC Method in the Simultaneous Analysis of 6-SHO and 6-GIN in Ginger Rhizome Extract, Commercial Ginger Powder, Capsules, and Ginger Teas

The obtained samples of traditional and ultrasonication-assisted extracts of ginger rhizome, commercial ginger powder, capsules, and ginger teas were applied on RP-HPTLC plates, and their chromatograms were obtained under the same experimental conditions and procedures as described for the simultaneous quantification of standard 6-SHO and 6-GIN. The HPTLC area of 6-SHO and 6-GIN in all studied test samples was recorded in three replicates (n = 3). The amount of 6-SHO and 6-GIN in all studied test samples was computed using the respective CC of 6-SHO and 6-GIN.

2.10. Statistical Analysis

All the values are expressed as the mean ± SD of three or six replicates. The statistical analysis was carried out by applying Dunnett's test, using GraphPad Prism software (version 6, GraphPad, San Diego, CA, USA). This analysis was performed at 5% level of significance.

3. Results and Discussion

3.1. Method Development

Based on available reports in literature, it was found that no green RP-HPTLC method reported for the simultaneous determination of 6-SHO and 6-GIN in plant extracts, food products, and formulations. Hence, this study was carried out to develop and validate a green RP-HPTLC method for the simultaneous quantification of 6-SHO and 6-GIN. In this study, the green mobile phase was obtained by the simple mixture of ethanol and water (green solvents) in comparison to the normal HPTLC method. The application of RP-HPTLC methods offer several advantages over normal phase HPTLC methods [37,40]. In addition, green RP-HPTLC methods for the simultaneous determination of 6-SHO and 6-GIN also reduce the toxicity of the proposed method to the environment [35,36].

In the proposed research, different compositions of ethanol and water, such as 5:5 (%, *v/v*), 6:4 (%, *v/v*), 7:3 (%, *v/v*), and 6.5:3.5 (%, *v/v*), were investigated as the green mobile phases for the development of a suitable band for the simultaneous RP-HPTLC-densitometry analysis of 6-SHO and 6-GIN. All studied mobile phases were developed using chamber saturation conditions.

From the results, it was found that the mixtures of ethanol and water, such as 5:5 (%, *v/v*), 6:4 (%, *v/v*), and 7:3 (%, *v/v*), presented poor densitometry peaks of 6-SHO and 6-GIN with poor symmetry. When the mixture of ethanol and water, such as 6.5:3.5 (%, *v/v*), was investigated, it was found that this combination presented well-resolved, symmetrical, and compact densitometry peaks of 6-SHO at $R_f = 0.36 \pm 0.01$ and of 6-GIN at $R_f = 0.53 \pm 0.01$ (Figure 2).

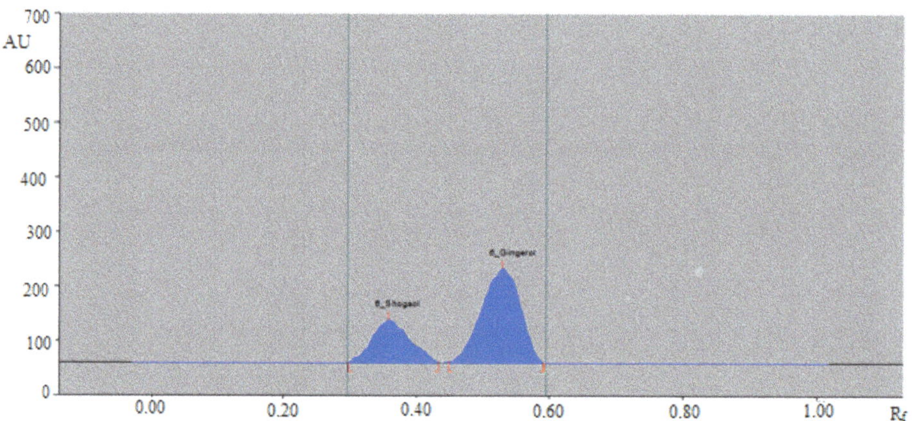

Figure 2. High-performance thin layer chromatography (HPTLC)-densitogram of standard 6-SHO and 6-GIN.

Based on these results, the mixture of ethanol: water 6.5:3.5 (%, *v/v*) was optimized as the green mobile phase for the simultaneous analysis of 6-SHO and 6-GIN in traditional and ultrasonication-assisted extracts of ginger rhizome, commercial ginger powder, capsules, and ginger teas. The band spectra for 6-SHO and 6-GIN were obtained in the densitometry mode, and the maximum response under reflectance/absorbance mode was obtained at the wavelength (λ_{max}) = 200 nm for 6-SHO and 6-GIN. Therefore, all simultaneous analyses of 6-SHO and 6-GIN were performed at $\lambda_{max} = 200$.

3.2. Method Validation

Different validation parameters for the simultaneous analysis of 6-SHO and 6-GIN were evaluated based on ICH guidelines [39]. The results for linear regression analysis of CCs of 6-SHO and 6-GIN are tabulated in Table 1.

Table 1. Linear regression analysis data for the calibration curve (CC) of 6-shogaol (6-SHO) and 6-gingerol (6-GIN) for the green reversed phase high-performance thin layer chromatography (RP-HPTLC) method (mean ± SD; $n = 6$).

Parameters	6-SHO	6-GIN
Linearity range (ng/band)	100–700	50–600
Regression equation	Y = 9.04x + 53.78	Y = 13.22x + 693.37
R^2	0.9988	0.9995
Slope ± SD	9.04 ± 0.54	13.22 ± 1.08
Intercept ± SD	53.78 ± 2.86	693.37 ± 15.75
Standard error of slope	0.22	0.44
Standard error of intercept	1.16	6.42
95% confidence interval of slope	8.09–9.99	11.33–15.12
95% confidence interval of intercept	48.76–58.81	665.70–721.03
LOD ± SD (ng/band)	33.65 ± 0.84	16.84 ± 0.36
LOQ ± SD (ng/band)	100.95 ± 2.52	50.52 ± 1.08

LOD: limit of detection and LOQ: limit of quantification.

The CC of 6-SHO and 6-GIN was found to be linear in the range of 100–700 and 50–600 ng/band, respectively. The results indicated a good linear relationship between the concentration and response of 6-SHO and 6-GIN. The value of determination coefficient (R^2) for 6-SHO and 6-GIN was computed as 0.9988 and 0.9995, respectively. The obtained R^2 values for 6-SHO and 6-GIN were significant ($p < 0.05$). Overall, the results of linear regression analysis indicated that a green RP-HPTLC method was linear for the simultaneous quantification of 6-SHO and 6-GIN.

Method accuracy for 6-SHO and 6-GIN was computed as % recovery, and results are summarized in Table 2.

Table 2. Accuracy data of 6-SHO and 6-GIN for the green RP-HPTLC method (mean ± SD; $n = 6$).

Excess Drug Added to Analyte (%)	Theoretical Content (ng)	Conc. Found (ng) ± SD	Recovery (%)	RSD (%)
		6-SHO		
0	100	98.8 ± 1.4	98.8	1.4
50	150	148.5 ± 2.0	99	1.3
100	200	202.4 ± 2.6	101.2	1.2
150	250	254.1 ± 2.8	101.6	1.1
		6-GIN		
0	100	101.5 ± 1.4	101.5	1.3
50	150	152.1 ± 1.9	101.4	1.2
100	200	198.3 ± 2.3	99.1	1.1
150	250	247.6 ± 2.6	99	1

RSD: relative standard deviation.

The % recoveries of 6-SHO and 6-GIN after spiking an extra 0–150% were obtained as 98.8–101.6 and 99.0–101.5%, respectively, using a green RP-HPTLC method. The percent of relative standard deviation (% RSD) values in the recovery of 6-SHO and 6-GIN were obtained as 1.10–1.46 and 1.07–1.39%, respectively. The obtained % recoveries within the limit of 100 ± 2% for 6-SHO and 6-GIN indicated that the green RP-HPTLC-densitometry method was accurate for the simultaneous quantification of 6-SHO and 6-GIN. Method precision for 6-SHO and 6-GIN was computed as % RSD, and results are summarized in Table 3.

Table 3. Precision data of 6-SHO and 6-GIN for the green RP-HPTLC method (mean ± SD; n = 6).

Conc. (ng/band)	Repeatability (Intraday Precision)			Intermediate Precision (Inter-Day)		
	Area ± SD	Standard Error	RSD (%)	Area ± SD	Standard Error	RSD (%)
			6-SHO			
100	956 ± 15	6.1	1.5	963 ± 16	6.5	1.6
150	1402 ± 17	7.1	1.2	1428 ± 17	7.2	1.2
200	1912 ± 18	7.5	0.9	1888 ± 16	6.5	0.8
250	2396 ± 20	8.4	0.8	2363 ± 18	7.4	0.7
			6-GIN			
100	2098 ± 21	8.7	1	2115 ± 21	8.9	1
150	2618 ± 24	9.8	0.9	2686 ± 25	10.2	0.9
200	3412 ± 27	11.3	0.8	3378 ± 25	10.5	0.7
250	4156 ± 32	13.2	0.7	4082 ± 29	12.1	0.7

The % RSD values of 6-SHO and 6-GIN for the intraday precision were calculated as 0.8–1.5 and 0.7–1.0%, respectively. The % RSD values of 6-SHO and 6-GIN for inter-day precision were obtained as 0.7–1.6 and 0.7–1.0%, respectively. The obtained values of % RSD for 6-SHO and 6-GIN within the range of ±2% showed that the green RP-HPTLC method was precise for the simultaneous quantification of 6-SHO and 6-GIN.

Robustness for 6-SHO and 6-GIN was determined by introducing minor modification in the composition of mobile phase, and results are summarized in Table 4. The errors (% RSD) for 6-SHO and 6-GIN after introducing minor modification in the composition of mobile phase were obtained as 0.7–0.8 and 0.6–0.8%, respectively. The small variations in R_f values and lower % RSD values suggested that the green RP-HPTLC method was robust for the simultaneous determination of 6-SHO and 6-GIN.

Table 4. Robustness data of 6-SHO and 6-GIN for the green RP-HPTLC method (mean ± SD; n = 6).

Conc. (ng/band)	Mobile Phase Composition (Ethanol:Water)			Results		
	Original	Used		Area ± SD	% RSD	R_f
			6-SHO			
		6.7:3.3	0.2	1503 ± 13	0.8	0.35
150	6.5:3.5	6.5:3.5	0	1432 ± 11	0.8	0.36
		6.3:3.7	−0.2	1342 ± 9	0.7	0.37
			6-GIN			
		Mobile phase composition (ethanol:water)				
		6.7:3.3	0.2	2753 ± 22	0.8	0.52
150	6.5:3.5	6.5:3.5	0	2654 ± 19	0.7	0.53
		6.3:3.7	−0.2	2513 ± 15	0.6	0.54

The method sensitivity for 6-SHO and 6-GIN was evaluated in terms of LOD and LOQ, and results are summarized in Table 1. The LOD and LOQ values for 6-SHO were calculated as 33.65 ± 0.84 and 100.95 ± 2.52 ng/band, respectively. However, the LOD and LOQ values for 6-GIN were calculated as 16.84 ± 0.36 and 50.52 ± 1.08 ng/band, respectively. The obtained values of LOD and LOQ indicated that the green RP-HPTLC method was sensitive enough for the simultaneous determination of 6-SHO and 6-GIN.

Method specificity and the peak purity of 6-SHO and 6-GIN were evaluated by comparing the overlaid UV-absorption spectra of 6-SHO and 6-GIN in ginger rhizome extract, commercial powder, capsules, and ginger root teas with that of the standard. The overlaid UV spectra of standard 6-SHO and 6-GIN and 6-SHO and 6-GIN in traditional and ultrasonication-assisted extracts of ginger rhizome, commercial ginger powder, capsules, and ginger teas are shown in Figure 3. Although the UV spectra of 6-SHO and 6-GIN in ultrasonication-assisted extract of ginger powder was quite different, their maximum responses were recorded as λ_{max} = 200 nm. The quite different spectra of 6-SHO and 6-GIN in ultrasonication-assisted extract of ginger powder could be due to the presence of other compounds. The maximum densitometry response of 6-SHO and 6-GIN in standards and traditional and ultrasonication-assisted extracts of ginger rhizome, commercial ginger powder, capsules, and ginger root teas were found at λ_{max} = 200 nm under the reflectance/absorbance mode. The similar UV-absorption spectra, R_f values, and λ_{max} of 6-GIN and 6-SHO in standard and traditional and ultrasonication-assisted extracts of ginger rhizome, commercial ginger powder, capsules, and ginger root teas suggested the method specificity for the simultaneous determination of 6-SHO and 6-GIN.

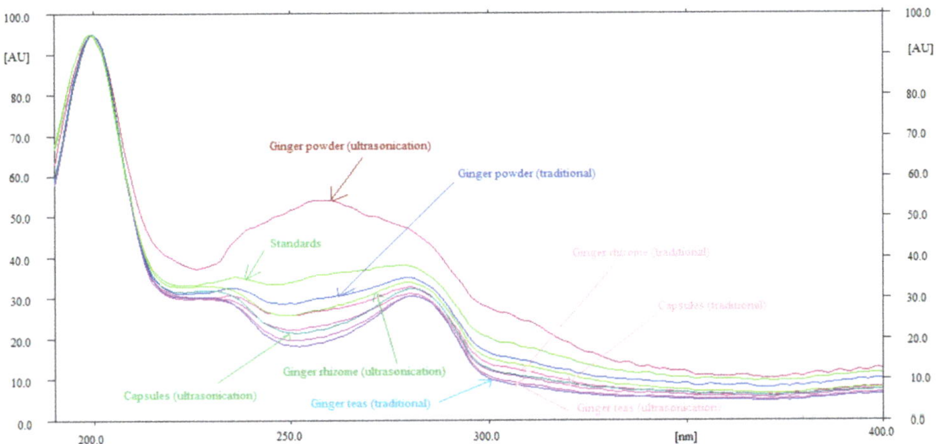

Figure 3. Overlaid ultra-violet (UV) absorption spectra of standards (6-SHO and 6-GIN), ginger rhizome, ginger powder, commercial capsules, and ginger teas extracted from traditional and ultrasonication methods.

3.3. Application of the Green RP-HPTLC Method in Simultaneous Analysis of 6-SHO and 6-GIN in Ginger Rhizome Eextract, Commercial Ginger Powder, Capsules, and Ginger Teas

A green RP-HPTLC method could be an alternative approach of conventional HPTLC techniques for the simultaneous determination of 6-SHO and 6-GIN in traditional and ultrasonication-assisted extracts of ginger rhizome, commercial ginger powder, capsules, and ginger teas. The densitometry peaks of 6-SHO and 6-GIN from traditional and ultrasonication-assisted extracts of ginger rhizome, commercial ginger powder, capsules, and ginger teas were verified by comparing their single TLC spot at R_f = 0.36 ± 0.01 for 6-SHO and R_f = 0.53 ± 0.01 for 6-GIN with that of standards 6-SHO and 6-GIN. The representative HPTLC densitogram of 6-SHO and 6-GIN in traditional ginger rhizome extracts is shown in Figure 4, which shows similar peaks of 6-SHO and 6-GIN to that of standard 6-SHO and 6-GIN. In addition, seven additional peaks were also recorded in the traditional ginger rhizome extract.

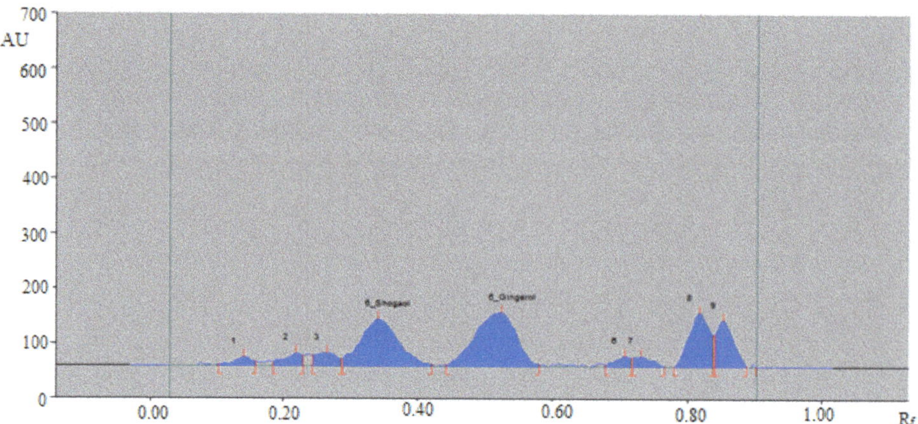

Figure 4. HPTLC-densitogram of 6-SHO and 6-GIN in traditional ginger rhizome extract.

The representative HPTLC densitogram of 6-SHO and 6-GIN in traditional commercial ginger powder extract is shown in Figure 5, which also shows similar peaks of 6-SHO and 6-GIN to that of standard 6-SHO and 6-GIN. In addition, three additional peaks were recorded in traditional commercial ginger powder extract.

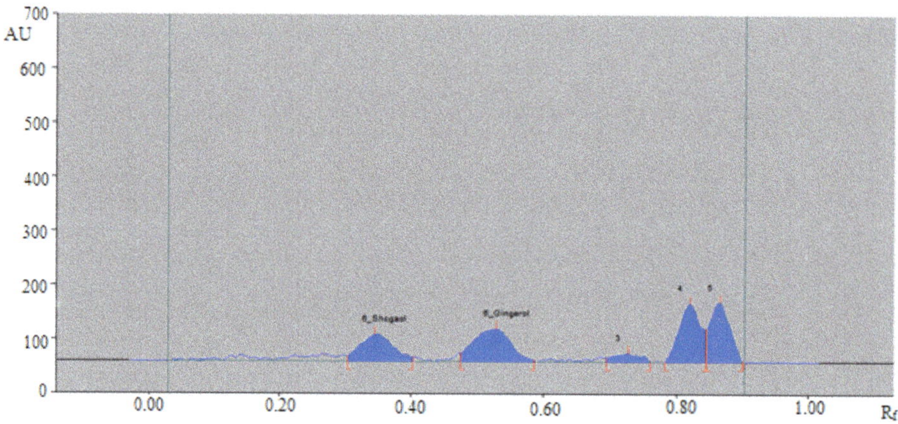

Figure 5. HPTLC-densitogram of 6-SHO and 6-GIN in traditional commercial ginger powder extract.

The representative HPTLC densitogram of 6-SHO and 6-GIN in traditional commercial capsule extract is shown in Figure 6, which also suggests similar peaks of 6-SHO and 6-GIN to that of standard 6-SHO and 6-GIN. In addition, three additional peaks were found in the traditional commercial capsule powder extract.

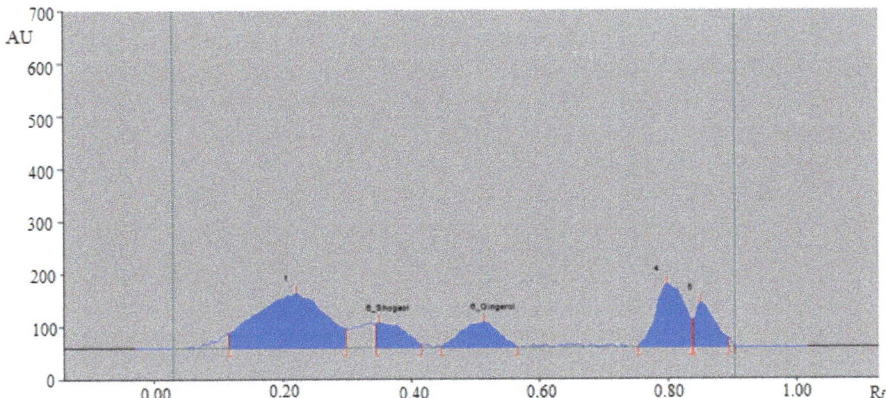

Figure 6. HPTLC-densitogram of 6-SHO and 6-GIN in traditional commercial capsules extract.

The presence of additional peaks in all studied traditional and ultrasonication-assisted extracts indicated that the green RP-HPTLC method could be successfully utilized for the simultaneous determination of 6-SHO and 6-GIN in the presence of other compounds or impurities. The amount of 6-SHO and 6-GIN in traditional and ultrasonication-assisted extracts of ginger rhizome, commercial ginger powder, capsules, and ginger teas was determined by the CCs of 6-SHO and 6-GIN. The amount of 6-SHO in traditional extracts of ginger rhizome, commercial ginger powder, capsules, and ginger teas was found as 12.1 ± 0.8, 17.9 ± 0.9, 10.5 ± 0.4, and 9.6 ± 0.3 mg/g of extract, respectively. However, the amount of 6-SHO in ultrasonication-assisted extracts of ginger rhizome, commercial ginger powder, capsules, and ginger teas was found as 14.6 ± 0.7, 19.7 ± 1.0, 11.6 ± 0.4, and 10.7 ± 0.4 mg/g of extract, respectively. The amount of 6-GIN in traditional extracts of ginger rhizome, commercial ginger powder, capsules, and ginger teas was recorded as 10.2 ± 0.6, 15.1 ± 0.8, 7.3 ± 0.2, and 6.2 ± 0.2 mg/g of extract, respectively. Meanwhile, the amount of 6-GIN in ultrasonication-assisted extracts of ginger rhizome, commercial ginger powder, capsules, and ginger teas was found as 12.7 ± 0.7, 17.8 ± 0.8, 8.8 ± 0.3, and 7.9 ± 0.3 mg/g of extract, respectively. In general, the amount of 6-SHO and 6-GIN was found to be significantly higher in traditional and ultrasonication-assisted extracts of commercial ginger powder and ginger rhizome compared with traditional and ultrasonication-assisted extracts of commercial capsules and ginger teas ($p < 0.05$). In addition, the amount of 6-SHO and 6-GIN in ultrasonication-assisted extracts was higher than traditional extracts. The amount of 6-SHO and 6-GIN in ultrasonication-assisted extracts of commercial ginger powder and ginger rhizome was significantly higher than their traditional extracts ($p < 0.05$). However, the amount of 6-SHO and 6-GIN in ultrasonication-assisted extracts of commercial capsules and ginger teas was not significant to their traditional extracts ($p > 0.05$). Based on these results, the ultrasonication method for the extraction of 6-SHO and 6-GIN in different ginger products was considered superior over the traditional method of extraction. Overall, these results suggested that the green RP-HPTLC method can be successfully applied in the simultaneous determination of 6-SHO and 6-GIN in a wide variety of food and pharmaceutical products containing 6-SHO and 6-GIN as the active constituents.

3.4. Literature Comparison

The green RP-HPTLC method for the simultaneous determination of 6-SHO and 6-GIN was compared with different analytical methods reported for the simultaneous determination of 6-SHO and 6-GIN. The results of comparison are tabulated in Table 5.

Different validation parameters, such as linearity range, accuracy, and precision of the green RP-HPTLC method, were compared with literature methods. The linearity range, accuracy, and precision of the reported LC-MS method for the simultaneous determination of 6-SHO and

6-GIN were found to be much inferior to the green RP-HPTLC method [29]. In addition, the linearity range, accuracy, and precision of various reported HPLC methods for the simultaneous determination of 6-SHO and 6-GIN were also found to be inferior to the green RP-HPTLC method [24,25,27]. HPTLC methods have not been reported for the simultaneous determination of 6-SHO and 6-GIN or 6-SHO alone. However, various HPTLC methods have been reported for the determination of 6-GIN alone or in combination with other compounds [20–23]. Therefore, the green RP-HPTLC method was also compared with reported HPTLC methods for the determination of 6-GIN alone. The results of this comparative analysis are tabulated in Table 6. The accuracy and precision for the determination of 6-GIN of most of the reported HPTLC methods were found within the limit of ICH guidelines [36]. However, the accuracy of the HPTLC method reported by Khan et al. was outside the limit of the ICH guidelines [23]. The accuracy of the present HPLTLC method was superior to those reported by Khan et al. [23]. However, there were not many differences found in the accuracy or precision of the present green RP-HPTLC method with other reported methods [20–22,24].

Table 5. Comparison of the current green RP-HPTLC method with reported analytical methods for the simultaneous determination of 6-SHO and 6-GIN.

Analytical Method	Compound						Ref.
	6-SHO			6-GIN			
	Linearity Range	Accuracy (% Recovery)	Precision (% RSD)	Linearity Range	Accuracy (% Recovery)	Precision (% RSD)	
LC-MS	1–40 (µg/mL)	83–110	2.0–4.0	5.5–220 (µg/mL)	87–100	2.0–8.0	[29]
HPLC	1–5 (µg/mL)	97.8–100.8	0.4–1.5	1.0–5.4 (µg/mL)	97.8–100.8	0.4–1.5	[24]
HPLC	6–18 (µg/mL)	84.7–92.8	3.0	20–60 (µg/mL)	91.5–102.3	3.4	[25]
HPLC	10–250 (µg/mL)	99.8–101.1	0.2–1.6	10–250 (µg/mL)	99.3–99.7	0.4–1.5	[27]
HPTLC	100–700 (ng/band)	98.8–101.6	0.7–1.6	50–600 (ng/band)	99.0–101.5	0.7–1.0	Present work

Table 6. Comparison of the current green RP-HPTLC method with reported HPTLC methods for the determination of 6-GIN.

Analytical Method	Linearity Range (ng/Band)	Accuracy (% Recovery)	Precision (% RSD)	Ref.
HPTLC	50–1000	98.1–98.8	0.6–1.2	[20]
HPTLC	140–840	95.6–101.4	0.7–1.4	[23]
HPTLC	150–900	98.2–99.6	0.0–0.1	[21]
HPTLC	100–1400	99.7–100.1	1.0–1.4	[22]
HPTLC	50–500	98.2–99.2	0.4–1.6	[24]
HPTLC	50–600	99.0–101.5	0.7–1.0	Present work

On the other hand, the linearity range (50–1000 ng/band) of one reported HPTLC method was wider than the present RP-HPTLC method (50–600 ng/band) with similar lower limit detections [20]. The linearity range (140–840 ng/band) of the other reported HPTLC method was slightly wider than the present RP-HPTLC method (50–600 ng/band), but its lower limit detections were much higher than the present HPTLC method [23]. The linearity range (50–500 ng/band) of the other reported HPTLC method was slightly narrower than the present RP-HPTLC method (50–600 ng/band), with similar lower limit detections [24]. The lower limit of detections of the other two HPTLC methods were much lower than the present RP-HPTLC method [21,22]. Overall, both reported methods, as well as the present HPTLC method, were found to be suitable for the determination of 6-GIN. However, the reported HPTLC methods were not utilized for the simultaneous determination of 6-SHO and 6-GIN. Overall, the present RP-HPTLC method for the simultaneous determination of 6-SHO and 6-GIN was found to be more simple, accurate, precise, cost-effective, and sensitive than reported analytical methods.

4. Conclusions

Due to the lack of green RP-HPTLC methods for the simultaneous determination of 6-SHO and 6-GIN in literature, the objective of this study was to develop and validate a green RP-HPTLC method for the simultaneous determination of 6-SHO and 6-GIN in traditional and ultrasonication-assisted extracts of ginger rhizome, commercial ginger powder, capsules, and ginger teas for the first time. The green RP-HPTLC is simple, accurate, precise, robust, sensitive, and specific for the simultaneous determination of 6-SHO and 6-GIN. The amount of 6-SHO and 6-GIN was found to be higher in ultrasonication-assisted extracts of ginger rhizome, commercial ginger powder, capsules, and ginger teas than their respective traditional extracts. Based on these results, ultrasonication has been proposed as the preferred method for the extraction of 6-SHO and 6-GIN from ginger rhizome, commercial ginger powder, capsules, and ginger teas over the traditional method of extraction. Overall, the results of this study suggested that the green RP-HPTLC method can be successfully utilized in the simultaneous analysis of 6-SHO and 6-GIN in the real samples of ginger food and pharmaceutical products having 6-SHO and 6-GIN as the active constituents.

Author Contributions: Conceptualization, supervision, P.A.; methodology, A.I.F., P.A., S.A.R., F.S., and H.S.Y.; validation, A.I.F., P.A., and F.S.; data curation, A.I.F., H.S.Y., and P.A.; funding acquisition, A.I.F.; project administration, A.I.F.; software, P.A., H.S.Y., and F.S.; writing—original draft, F.S.; writing—review and editing, A.I.F., H.S.Y., S.A.R., and P.A. All authors have read and agreed to the published version of the manuscript.

Funding: This project was funded by Support Research Projects with the Global Partnership Program, Deanship of Scientific Research, Prince Sattam Bin Abdulaziz University, Al-Kharj, KSA (Grant No. 2020/03/17097). The article processing charge (APC) was also funded by the Deanship of Scientific Research, Prince Sattam Bin Abdulaziz University, Al-Kharj, KSA.

Acknowledgments: The authors would like to express their gratitude to the Deanship of Scientific Research of Prince Sattam Bin Abdulaziz University, Al-Kharj for their assistance.

Conflicts of Interest: The authors declare that they have no conflict of interests associated with this manuscript.

References

1. Pawar, N.; Pai, S.; Nimbalkar, M.; Dixit, G. RP-HPLC analysis of phenolic antioxidant compound 6-gingerol from different ginger cultivars. *Food Chem.* **2011**, *126*, 1330–1336. [CrossRef]

2. Gaikwad, D.A.; Shinde, S.K.; Kawade, A.V.; Jadhav, S.J.; Gadhave, M.V. Isolation and standardization of gingerol from ginger rhizome by using TLC. HPLC, and identification tests. *Pharm. Innov.* **2017**, *6*, 179–182.

3. Sekiwa, Y.; Kubota, K.; Kobayashi, A. Isolation of novel glucosides related to gingerdiol from ginger and their antioxidative activities. *J. Agric. Food Chem.* **2000**, *48*, 373–377. [CrossRef]

4. Young, H.Y.; Luo, Y.L.; Cheng, H.Y.; Hsieh, W.C.; Liao, J.C.; Peng, W.H. Analgesic and anti-inflammatory activities of [6]-gingerol. *J. Ethnopharmacol.* **2005**, *96*, 207–210. [CrossRef]

5. Wei, Q.Y.; Ma, J.P.; Cai, Y.J.; Yang, L.; Liu, Z.L. Cytotoxic and apoptotic activities of diarylheptanoids and gingerol-related compounds from the rhizome of Chinese ginger. *J. Ethnopharmacol.* **2005**, *102*, 177–184. [CrossRef] [PubMed]

6. Shukla, Y.; Singh, M. Cancer preventive properties of ginger: A brief review. *Food Chem. Toxicol.* **2007**, *45*, 683–690. [CrossRef] [PubMed]

7. Marx, W.; McKavanagh, D.; McCarthy, A.L.; Bird, R.; Ried, K.; Chan, A.; Isenring, L. The effect of ginger (*Zingiber officinale*) on platelet aggregation: A systematic literature review. *PLoS ONE* **2015**, *10*, e141119.

8. Shinde, S.K.; Grampurothi, N.D.; Banerjee, S.K.; Jadhav, S.L.; Gaikwad, D.D. Development and validation of UV spectroscopic method for the quick estimation of gingerol from *Zingiber officinale* rhizome extract. *Int. Res. J. Pharm.* **2012**, *3*, 234–237.

9. Young, H.Y.; Chiang, C.T.; Huang, Y.L.; Pan, F.P.; Chen, G.L. Analytical and stability studies of ginger preparations. *J. Food Drug Anal.* **2002**, *10*, 149–153.

10. Abu-Yousef, I.A.; Gunasekar, C.; Dghaim, R.; Abdo, N.; Narasimhan, S. Simplified HPLC method for identification of gingerol and mengiferin in herbal extracts. *Eur. J. Sci. Res.* **2011**, *66*, 21–88.

11. Mishra, A.P.; Saklani, S.; Chandra, S. Estimation of gingerol contents in different brand samples of ginger powder and their anti-oxidant activity. *Rec. Res. Sci. Technol.* **2013**, *5*, 54–59.

12. Chen, Y.; Zhang, C.; Zhang, M.; Fu, X. Assay of 6-gingerol in CO_2 supercritical fluid extracts of ginger and evaluation of its sustained release from a transdermal delivery system across rat skin. *Pharmazie* **2014**, *69*, 506–511. [PubMed]

13. Cho, S.; Lee, D.G.; Lee, S.; Chae, S.; Lee, S. Analysis of the 6-gingerol content in *Zingiber* spp. and their commercial foods using HPLC. *J. Appl. Biol. Chem.* **2015**, *58*, 377–381. [CrossRef]

14. Cafino, E.J.V.; Lirazan, M.B.; Marfori, E.C. A simple HPLC method for the analysis of [6]-gingerol produced by multiple shoot culture of ginger (*Zingiber officinale*). *Int. J. Pharmacogn. Phytochem.* **2016**, *8*, 38–42.

15. Yang, H.Y.; Ma, J.Y.; Weon, J.B.; Lee, B.; Ma, C.J. Qualitative and quantitative simultaneous determination of six marker compounds in Shoshiho-tang by HPLC-DAD-ESI-MS. *Arch. Pharm. Res.* **2012**, *10*, 1785–1791. [CrossRef]

16. Zeng, S.L.; Liu, X.G.; Lai, C.J.S.; Liu, E.H.; Li, P. Diagnostic ion filtering strategy for chemical characterization of Guge Fengtong tablet with high-performance liquid chromatography coupled with electrospray ionization quadrupole time-of-flight tandem mass spectrometry. *Chin. J. Nat. Med.* **2015**, *13*, 390–400. [CrossRef]

17. Ngamdokmai, N.; Waranuch, N.; Chootip, K.; Neungchamnong, N.; Ingkaninan, K. HPLC-QTOF-MS method for quantitative determination of active compounds in an anti-cellulite herbal compress. *Songklanakarin J. Sci. Technol.* **2017**, *39*, 463–470.

18. Nishidodno, Y.; Saifuddin, A.; Nishizawa, M.; Fujita, T.; Nakamoto, M.; Tanaka, K. Identification of chemical constituents in ginger (*Zingiber officinale*) responsible for thermogenesis. *Nat. Prod. Commun.* **2018**, *13*, 869–873.

19. Dong, Y.Z.; Liu, Z.L.; Liu, Y.Y.; Song, Z.Q.; Guo, N.; Wang, C.; Ning, Z.C.; Ma, X.L.; Lu, A.P. Quality control of the Fuzi Lizhong pill through simultaneous determination of 16 major bioactive constituents by RRLC-MS-MS. *J. Chromatogr. Sci.* **2018**, *56*, 541–554. [CrossRef]

20. Alqasoumi, S.I. Quantification of 6-gingerol in *Zingiber officinale* extract, ginger-containing dietary supplements, teas and commercial creams by validated HPTLC densitometry. *Fabad J. Pharm. Sci.* **2009**, *34*, 33–42.

21. Jain, P.S.; Tatiya, A.U.; Bagul, S.A.; Surana, S.J. Development and validation of a method for densitometric analysis of 6-gingerol in herbal extracts and polyherbal formulation. *J. Anal. Bioanal. Tech.* **2011**, *2*, 1–3. [CrossRef]

22. Kumar, K.S.; Manasa, B.; Rahman, K.; Sudhakar, B. Development and validation of HPTLC method for estimation of 6-gingerol in herbal formulations and extracts. *Int. J. Pharm. Sci. Res.* **2012**, *3*, 3762–3765.

23. Khan, I.; Pandotra, P.; Gupta, A.P.; Sharma, R.; Gupta, B.D.; Dhar, J.K.; Ram, G.; Bedi, Y.S.; Gupta, S. RP-thin layer chromatographic method for the quantification of three gingerol homologs of ultrasonic-assisted fresh rhizome extracts of *Zingiber officinale* collected from North Western Himalayas. *J. Sep. Sci.* **2010**, *33*, 558–563. [CrossRef]

24. Alam, P. Densitometric HPTLC analysis 8-gingerol in *Zingiber officinale* extract and ginger-containing dietary supplements, teas and commercial creams. *Asian Pac. J. Trop. Biomed.* **2013**, *3*, 634–638. [CrossRef]

25. Pandotra, P.; Sharma, R.; Datt, P.; Kushwaha, M.; Gupta, A.K.; Gupta, S. Ultrasound-assisted extraction and fast chromolithic method development, validation and system suitability analysis for 6-, 8-, 10-gingerols and 6-shogaol in rhizome of *Zingiber officinale* by liquid chromatography-diode array detection. *Int. J. Green Pharm.* **2013**, *7*, 189–195.

26. Kajsongkram, T.; Rotamporn, S.; Limbunruang, S.; Thubthimthed, S. Development and validation of a HPLC method for 6-gingerol and 6-shogaol in joint pain relief gel containing ginger (*Zingiber officinale*). *Int. J. Med. Health. Sci.* **2015**, *9*, 813–817.

27. Ali, A.M.A.; El-Nour, M.E.M.; Mohammad, O.; Yagi, S.M. In vitro anti-inflammatory activity of ginger (*Zingiber officinale* Rosc.) rhizome, callus and callus treated with some elicitors. *J. Med. Plant Res.* **2019**, *13*, 227–235.

28. Antony, B.; Benny, B.; Reshma, M. Validation of reversed-phase high-performance liquid chromatography method for simultaneous determination of 6-, 8-, 10-gingerols and 6-shogaol from ginger extracts. *Int. J. Pharm. Phar. Sci.* **2020**, *12*, 67–70. [CrossRef]

29. Zick, S.M.; Ruffin, M.T.; Djuric, Z.; Normolle, D.; Brenner, D.E. Quantification of 6-, 8-, 10-gingerols and 6-shogaol in human plasma by high-performance liquid chromatography with electrochemical detection. *Int. J. Biomed. Sci.* **2010**, *6*, 233–240.

30. Lee, S.; Khoo, C.; Halstead, W.; Huynh, T.; Bensossan, A. Liquid chromatographic determination of 6-, 8-, 10-gingerol and 6-shogaol in ginger (*Zingiber officinale*) as the raw herb and dried aqueous extract. *J. AOAC Int.* **2007**, *90*, 1219–1226. [CrossRef]

31. Alam, P.; Ezzeldin, E.; Iqbal, M.; Anwer, M.K.; Mostafa, G.A.E.; Alqarni, M.H.; Foudah, A.I.; Shakeel, F. Ecofriendly densitometric RP-HPTLC method for determination of rivaroxaban in nanoparticle formulations using green solvents. *RSC Adv.* **2020**, *10*, 2133–2140. [CrossRef]

32. Alam, P.; Ezzeldin, E.; Iqbal, M.; Mostafa, G.A.E.; Anwer, M.K.; Alqarni, M.H.; Foudah, A.I.; Shakeel, F. Determination of delafloxacin in pharmaceutical formulations using a green RP-HPTLC and NP-HPTLC methods. *Antibiotics* **2020**, *9*, 359. [CrossRef]

33. Foudah, A.I.; Alam, P.; Anwer, M.K.; Yusufoglu, H.S.; Abdel-Kader, M.S.; Shakeel, F. A green RP-HPTLC-densitometry method for the determination of diosmin in pharmaceutical formulations. *Processes* **2020**, *8*, 817. [CrossRef]

34. Bhandari, P.; Kumar, N.; Gupta, A.P.; Singh, B.; Kaul, V.K. A rapid RP-HPTLC densitometry method for simultaneous determination of major flavonoids in important medicinal plants. *J. Sep. Sci.* **2007**, *30*, 2092–2096. [CrossRef]

35. Sharma, U.K.; Sharma, N.; Gupta, A.P.; Kumar, V.; Sinha, A.K. RP-HPTLC determination and validation of vanillin and related phenolic compounds in accelerated solvent extract of *Vanilla planifolia*. *J. Sep. Sci.* **2007**, *30*, 3174–3180. [CrossRef] [PubMed]

36. Al-Alamein, A.M.A.; Abd El-Rahman, M.K.; Abdel-Moety, E.M.; Fawaz, E.M. Green HPTLC-densitometric approach for simultaneous determination and impurity-profiling of ebastine and phenylephrine hydrochloride. *Microchem. J.* **2019**, *147*, 1097–1102. [CrossRef]

37. Rezk, M.R.; Monir, H.; Marzouk, H.M. Novel determination of a new antiviral combination; sofosbuvir and velpatasvir by high performance thin layer chromatographic method; application to real human samples. *Microchem. J.* **2019**, *146*, 828–834. [CrossRef]

38. Foudah, A.I.; Alam, P.; Shakeel, F.; Alqasoumi, S.I.; Alqarni, M.H.; Yusufoglu, H.S. Eco-friendly RP-HPTLC method for determination of valerenic acid in methanolic extract of Valeriana officinalis and commercial herbal products. *Lat. Am. J. Pharm.* **2020**, *39*, 420–424.

39. International Conference on Harmonization (ICH). *Q2 (R1): Validation of Analytical Procedures–Text and Methodology*; ICH: Geneva, Switzerland, 2005.

40. Ahmed, S.; Al-Rehaily, A.J.; Alam, P.; Alqahtani, A.S.; Hidayatullah, S.; Rehman, M.T.; Mothana, R.A.; Abbas, S.S.; Khan, M.U.; Khalid, J.M.; et al. Antidiabetic, antioxidant, molecular docking and HPTLC analysis of miquelianin isolated from *Euphorbia schimperi* C. Presl. *Saudi Pharm. J.* **2019**, *27*, 655–663. [CrossRef]

 *foods*

Article

NMR-Based Metabolomic Comparison of *Brassica oleracea* (Var. *italica*): Organic and Conventional Farming

Massimo Lucarini [1,*], Maria Enrica Di Cocco [2], Valeria Raguso [2], Flavia Milanetti [2], Alessandra Durazzo [1], Ginevra Lombardi-Boccia [1], Antonello Santini [3], Maurizio Delfini [2] and Fabio Sciubba [2,*]

1   CREA—Research Centre for Food and Nutrition, Via Ardeatina 546, 00178 Rome, Italy; alessandra.durazzo@crea.gov.it (A.D.); g.lombardiboccia@crea.gov.it (G.L.-B.)
2   Department of Chemistry, "Sapienza" University of Rome, Piazzale Aldo Moro 5, 00181 Rome, Italy; mariaenrica.dicocco@uniroma1.it (M.E.D.C.); valeria.raguso@gmail.com (V.R.); flavia.milanetti@uniroma1.it (F.M.); maurizio.delfini@uniroma1.it (M.D.)
3   Department of Pharmacy, University of Napoli Federico II, Via D. Montesano 49, 80131 Napoli, Italy; asantini@unina.it
*   Correspondence: massimo.lucarini@crea.gov.it (M.L.); fabio.sciubba@uniroma1.it (F.S.)

Received: 12 June 2020; Accepted: 15 July 2020; Published: 17 July 2020

Abstract: Brassicaceae family provides several crops which are worldwide known for their interesting phytochemical profiles, especially in terms of content of glucosinolates. These secondary metabolites show several beneficial effects toward consumers' health, and several studies have been conducted to identify cultivation factors affecting their content in crops. One of the agronomic practices which is attracting growing interest is the organic one, which consists in avoiding the use of mineral fertilizers as well as pesticides. The aim of this study is to define the metabolic profile of *Brassica oleracea* (var. *italica*) and to compare the samples grown using organic and conventional fertilization methods. The hydroalcoholic and organic extracts of the samples have been analyzed by NMR spectroscopy. Forty-seven metabolites belonging to the categories of organic acids, amino acids, carbohydrates, fatty acids, sterols, and other molecules have been identified. Thirty-seven metabolites have been quantified. Univariate and multivariate PCA analyses allowed to observe that the organic practice influenced the nitrogen transport, the carbohydrate metabolism, the glucosinolate content and the phenylpropanoid pathway in *B. oleracea* (var. *italica*).

Keywords: NMR; metabolomics; *Brassica oleracea* (var. *italica*); organic and conventional pratices; glucosinolates

1. Introduction

A healthy lifestyle is a combination of behaviors and one of the main aspects is the diet. It is recommended to eat as a minimum five portions of fruits and vegetables daily, thus reducing the risk of chronic disease and improve health outcomes [1]. Vegetables are not only a natural source of amino acids, minerals and vitamins, but they are rich in several secondary plant metabolites that can be subdivided into different groups depending on their chemical structure and functional properties [2]. Some of the secondary plant metabolites with nutraceutical properties [3–5] can be found ubiquitously in the entire plant kingdom, and thus, in all types of vegetables. On the other hand, the large and very diverse group of phenolic compounds or carotenoids as well as other secondary plant produced metabolites, are restricted to some botanical orders or families e.g., the glucosinolates (GLs), which are distributed mostly in the order of flowering plants *Brassicales*.

Brassicaceous vegetables belong to the order *Brassicale*s and most of them are members of the Brassicaceae family. About the 12% of the world-grown vegetables are *Brassica* vegetables [6], illustrating the great importance of this family. Two very common groups of *Brassicaceous* vegetables are the *Brassica oleracea* (e.g., broccoli, Brussels sprouts, white and red cabbage, cauliflower, collards, kale, and kohlrabi) and *Brassica rapa* (Chinese cabbage, pak choi, and turnips).

Brassicaceous vegetables contain vitamins C, E, and K, as well as folate, minerals, and dietary fiber. *Brassica* generally contains high amounts of vitamin C and can provide up to the 50% of the daily recommended dietary intake of this vitamin [7].

In addition to the phytochemicals, such as carotenoids and phenolic compounds, which occur in considerable amounts in some Brassica species [8–10], *Brassicaceous* vegetables are rich also of sulfur-containing compounds e.g., methylcysteinsulfoxide, and glucosinolate [11,12], which are responsible for the pungent and bitter taste or the spicy flavor of *Brassicaceous* vegetables [13,14].

One of the most relevant and interest biomolecules in *Brassicales* vegetables are the glucosinolates, due to their health-promoting properties in general, and cancer preventive properties in particular as substantiated by many studies on this topic [15–19].

Glucosinolates are stable secondary metabolite in plants and play a key role in the plant's defense system. In case of tissue injury (e.g., insect's damage), they are enzymatically decomposed by the endogenous enzyme myrosinase and, as a result, various degradation products, such as nitriles, epithionitriles, and/or isothiocyanates (ITCs) are released [20]. Isothiocyanates are associated with the pungency of these vegetables and have been shown to confer several beneficial effects [21–23], i.e., anti-cancerogenic [2,15,24–30], anti-inflammatory [31], as well as anti-diabetogenic [32,33] properties and beneficial health effects.

The composition of secondary metabolites strongly depends on factors such as: (i) pedoclimatic conditions of sampling site; (ii) harvesting time; (iii) plant genotype; (iv) agronomic practices.

Concerning the agronomic practices, the debate about the differences in nutritional properties between organic and conventional food is currently open, as shown by the consistent number of papers and reviews published in the last few years. Comparisons between organic and conventional cultivation methods have shown that organic practices make the plant more susceptible to attacks by pathogens and insects causing an overproduction of secondary metabolites (i.e., phenolic and GLs) in response to the biotic stress respect to conventional. Furthermore, soil fertilization is another factor that can influence the content of plants phytochemicals and the interactive effect depends on crop varieties, plant tissue considered and soil type. As instance, Jones et al. [34] found that nitrogen stress increased glucoraphanin, quercetin and kaempferol content in broccoli florets and decreased glucobrassicin content. The authors hypothesized that the limited nitrogen results in an increased availability of methionine for aliphatic glucosinolate production, while the opposite occurred with tryptophan used for the synthesis of indoyl GLs.

However, the metabolic pathways underlying the biosynthesis of secondary metabolites cannot be analyzed without also studying the whole plant metabolism, which is clearly influenced by soil nitrogen supply.

In recent years, Nuclear Magnetic Resonance (NMR) has emerged as one of the main analytical techniques used in metabolomics. NMR allows to analyze at the same time all the metabolites present in a sample with a single experiment and a minimum of pre-treatment. In fact, the technological advances have allowed to overcome the most important negative problem represented by the low intrinsic sensitivity of this spectroscopic technique. Furthermore, the most advanced two-dimensional techniques allow to identify the compounds present also in extremely complex mixtures, making possible a qualitative and quantitative analysis. For these reasons, NMR spectroscopy is nowadays among the main analytical techniques used in the metabolomics research; it has several advantages including a relatively high degree of reproducibility, easy-to-identify metabolites, high throughput, and non-destructive sample treatment.

On the basis of the metabolic profile obtained from NMR experiments, it has been possible to identify a wide range of metabolites with a single analysis allowing to evaluate various food characteristics, regarding quality, authentication, geographical origin, as well as secondary metabolites with potential nutraceutical properties [35–39].

The aim of this work has been to investigate the effects and interactions of cultivation, organic versus conventional, on the secondary metabolites profile in *B. oleracea* adding two soil fertilizer with different rate of utilization of nitrogen. The two broccoli theses were grown in the same pedoclimatic and soil conditions in order to observe the specific effect of cultivation.

2. Materials and Methods

2.1. Sampling

For this experiment *B. oleracea* (var. *italica*) cultivar Natalino plants were grown in the Roma (Italy) countryside, in the area of Fiano Romano (Lazio Region, Italy). Natalino variety was selected for its agronomic traits, such as crop robustness, yield stability and an excellent tolerance to freezing stress of both the plant and corymbs. The two cultivation sites were characterized by high similarity in terms of sun exposure and pedoclimatic conditions. The soil is classified as sandy loan with standard mineral dotation of macro and micro elements. The site area is characterized by the Mediterranean climate, with mean annual temperature of 20 °C and annual precipitation of 482 mm concentrated in the Autumn and Spring seasons.

Seedlings were transplanted 6 weeks after germination. Fertilizer were applied at transplanting. Plants were regularly irrigated with drip irrigation system.

Seedlings were transplanted in September and harvested 100 days later in December

Broccoli florets were harvested at the standard commercial ripening stage with characteristic inflorescence consisting of very tight, regular, compact and intense bright green florets with fully developed corymbs.

Broccoli were grown under organic and conventional agriculture and thirty (30) corymbs from different plants were sampled, fifteen (15) for each of the two types of cultivation.

No chemicals were used for the control of pests and phytopathological diseases in either conventional or organic cultivations.

Organic and conventional cultivation provided the same amount of nitrogen to the soil and fertilizer applied was urea and bovine manure as reported in Table 1.

Table 1. Fertilizer application on conventional and organic fields.

	Fertilizer Application (t/ha)	
	Urea	**Bovine Manure**
Conventional	0.2	15
Organic	0.0	28

After sampling, the samples were put in polyethylene bags at the collection site to avoid water loss, and sent in a refrigerated container to laboratory, where they were stored at −80 °C until further analysis.

2.2. Sample Preparation and Metabolites Extraction

Homogeneous portions of the vegetables (0.5 g of fresh weight) were frozen in liquid nitrogen, finely powdered and extracted according to the modified Bligh-Dyer protocol [40]. Each sample aliquot was placed in a mortar, ground in liquid nitrogen and added to a cold mixture composed of with methanol/chloroform/water in a 2:2:1 proportion. The samples were kept at +4 °C for 1 h and then centrifuged for 25 min at 4 °C at 10,000 rpm on an Itettich Zentrifugen centrifuge (Tuttlingen, Germany). This extraction procedure ensures that the metabolic profile does not change and that it is as close as possible to the desired analysis time point. This extraction procedure was employed also because it allows to separate low weight compounds on the basis of their polarity. The upper hydroalcoholic phase and the lower organic one were carefully separated, dried under nitrogen flux and stored at −80 °C until NMR analysis.

2.3. NMR Spectroscopy

The hydrophilic phase was resuspended in 0.6 mL of D_2O containing 3-(trimethylsilyl)-propionic-2,2,3,3-D_4 acid sodium salt (final concentration of TSP, 2 mM) as an internal chemical shift and concentration standard. The hydrophobic phase was resuspended in 0.6 $CDCl_3$ with hexamethyldisiloxane (final concentration of HMDS, 2 mM) as an internal standard. All solvents and standards were purchased from Sigma Aldrich (St. Louis, MO, USA).

All spectra were recorded at 298 K on a Bruker AVANCE III spectrometer operating at the proton frequency of 400.13 MHz and equipped with a multinuclear z-gradient inverse probehead (Bruker BioSpin, GMBH, Rheinstetten, Germany). The ^1H 1D spectra and 2D ^1H-^1H TOCSY, ^1H-^{13}C HSQC and ^1H-^{13}C HMBC were acquired employing previously used parameters [41]. The signals that could be clearly identified and had no overlap with neighboring resonances were integrated for each sample and quantification was performed by comparison of the signal integral with the reference signal, and quantities were expressed in mg/g of fresh weight.

2.4. Statistics

Univariate *t*-test analysis was performed with SigmaPlot 14.0 software (Systat Software Inc., San Jose, CA, USA). Multivariate PCA was performed on the data matrix of metabolite concentrations measured by NMR spectroscopy with the Unscrambler ver. 10.5 software (Camo Software AS, Oslo, Norway). Data were mean centered, since the variables with the largest response could dominate the PCA, and then autoscaled to equalize the importance of the variation of each variable.

3. Results and Discussion

Comprehensive metabolic profile analysis of *B. oleracea* var. *italica* was carried out by ^1H NMR spectroscopy of hydroalcoholic and chloroform extracts. The extracts of the two cultivation practices showed only quantitative differences and not qualitative ones (Figures S1 and S2 for hydroalcoholic and chloroform extracts, respectively). Resonance assignment was carried through bidimensional TOCSY (Figures S3 and S4), HSQC (Figures S5 and S6) and HMBC (Figures S7 and S8) experiments and confirmed by literature data [41,42]. A total of 47 metabolites were identified; 37 were quantified, and the ^1H chemical shifts, multiplicity and the ^{13}C chemical shifts are reported in Supplementary Table S1.

The signals that could be clearly identified and had no overlap with neighboring signals were integrated for each sample and quantification was performed by comparison of the signal integral with the reference one, and quantities were expressed in mg/g of fresh weight. The resulting data set was studied by univariate and multivariate statistical analysis tools for the evaluation of statistical differences between the two fertilization methods. The quantitative analysis is reported in Table 2 with the statistical significance assessed by student *t*-test.

Comparing the profile of *B. oleracea* fertilised with only manure with the one fertilised with both urea and manure, it is possible to observe an increase of aspartate, glycine, tyrosine, histidine, malate, sucrose, linolenic fatty acid, total choline, glucoraphanin, and glucobrassicin as well as a decrease of valine, isoleucine, threonine, glutamine, lysine, arginine, asparagine, phenylalanine, acetate, and fructose.

Table 2. Composition of *B. Oleracea* var. *italica* determined by NMR spectroscopy.

Molecule		Amount (mg/g)		Change
		Conventional	Organic	
Free Amino acids	Valine	0.248 ± 0.046	0.172 ± 0.016 **	↓
	Isoleucine	0.115 ± 0.023	0.074 ± 0.003 **	↓
	Leucine	0.124 ± 0.028	0.106 ± 0.008	
	Threonine	0.153 ± 0.021	0.122 ± 0.006 **	↓
	Alanine	0.307 ± 0.055	0.307 ± 0.024	
	Glutamate	1.349 ± 0.399	1.794 ± 0.245	
	Glutamine	0.729 ± 0.135	0.521 ± 0.028 **	↓
	Aspartate	0.626 ± 0.197	0.843 ± 0.066 *	↑
	Lysine	0.314 ± 0.066	0.254 ± 0.008 *	↓
	Arginine	2.852 ± 0.508	2.378 ± 0.113 *	↓
	Asparagine	1.151 ± 0.234	0.631 ± 0.056 **	↓
	Glycine	0.528 ± 0.147	0.702 ± 0.128 **	↑
	Tyrosine	0.076 ± 0.010	0.100 ± 0.008 **	↑
	Histidine	0.027 ± 0.007	0.056 ± 0.010 **	↑
	Phenylalanine	0.273 ± 0.067	0.159 ± 0.017 **	↓
Organic acids	Acetate	0.042 ± 0.009	0.026 ± 0.002 **	↓
	Malate	2.552 ± 0.452	3.186 ± 0.175 *	↑
	Pyruvate	0.490 ± 0.209	0.650 ± 0.111	
	Succinate	0.148 ± 0.085	0.101 ± 0.012	
	Fumarate	0.148 ± 0.069	0.089 ± 0.020	
	Formate	0.005 ± 0.001	0.004 ± 0.001	
Carbohydrates	Glucose	4.552 ± 1.077	5.451 ± 0.293	
	Fructose	1.634 ± 0.449	1.041 ± 0.083 **	↓
	Sucrose	2.380 ± 0.752	6.946 ± 0.487 **	↑
Lipids and sterols	β-Sitosterol	0.381 ± 0.067	0.352 ± 0.042	
	Campesterol	0.117 ± 0.043	0.112 ± 0.018	
	Stearic acid	1.849 ± 0.794	1.733 ± 0.271	
	Oleic acid	1.024 ± 0.474	1.163 ± 0.168	
	Linoleic acid	0.673 ± 0.156	0.691 ± 0.071	
	Linolenic acid	1.310 ± 0.252	1.663 ± 0.147 *	↑
	Monoacylglycerol	0.356 ± 0.049	0.367 ± 0.029	
	Triacylglygerol	0.305 ± 0.052	0.362 ± 0.028	
Miscellaneous	Choline	0.337 ± 0.058	0.480 ± 0.053 *	↑
	Glucoraphanin	0.565 ± 0.087	0.708 ± 0.027 **	↑
	Glucobrassicin	0.160 ± 0.052	0.449 ± 0.105 **	↑
	Trigonelline	0.026 ± 0.004	0.027 ± 0.001	
	Indole-3-carbinol	0.017 ± 0.007	0.017 ± 0.003	

Values are mean ± standard deviation (sample number $n = 15$). Level of significance: * $p < 0.05$, ** $p < 0.01$; ↑ and ↓ indicate a significative increase and decrease respectively compared to conventional fertilization.

To observe correlations among the quantified molecules, a PCA analysis was carried out on the whole data matrix, providing a model whose first 6 components explained 80% of the overall variance with the first component (PC1) accounting for 22% and the second one (PC2) for 20% of the overall variance as shown in Figure 1.

Analyzing the PCA score plot, while there is not a clear grouping of the samples according to the farming, it has been possible to observe that conventional samples were mainly at positive values of PC1, while most of the organic ones were at negative values. A *t*-test performed on the PC1 values of the samples indicated the observed difference between the samples to be significant ($p < 0.01$). The variables important for the discrimination could be determined studying the PC1 loading values (see Figure 2) and normalized loading values greater than 0.349 and lower than −0.349 were considered significant ($p < 0.05$) according to Pearson table for covariance significance.

PCA Scores

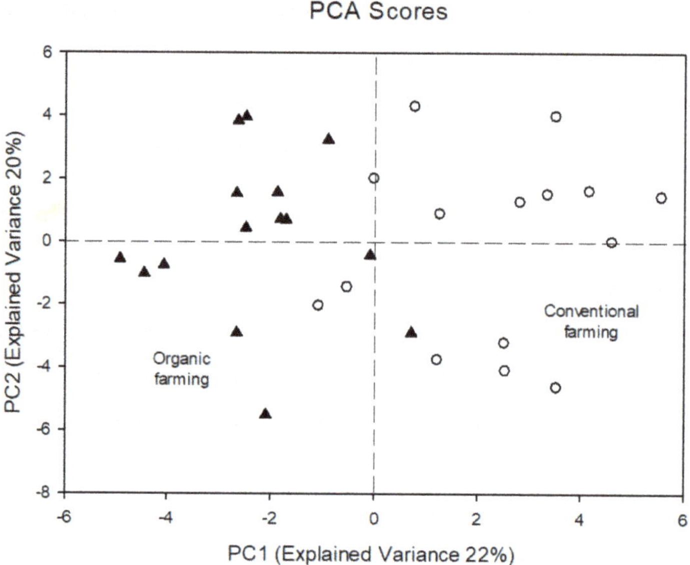

Figure 1. PCA score plot analysis (PC1 vs. PC2) of *Brassica oleracea* (var. *italica*) samples. Conventional samples are indicated by white circle and organic sample are indicated by black triangles (for the abbreviation used see Table S1).

PC1 Loadings

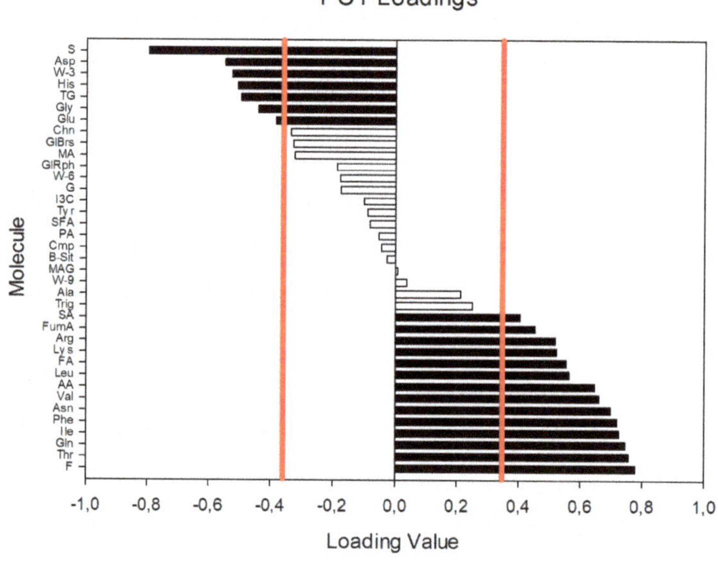

Figure 2. Normalized PC1 loading values. The red lined indicate the significance threshold and in black the variables with $p < 0.05$ are evidenced.

The molecules negatively correlating with PC1, and thus important for the definition of organic samples were sucrose, aspartate, linolenic fatty acid, histidine, glycerol of triglycerides, glycine, and glutamate, while the ones important for the description of conventional samples were fructose, threonine, glutamine, isoleucine, phenylalanine, asparagine, valine, acetate, leucine, formate, lysine, arginine, fumarate, and succinate.

The results of both univariate and multivariate analysis indicated the same molecules to be discriminant between conventional and organic farming. First of all, it is important to remember/underline that for this experimentation no pesticides and other treatments were employed in any plant, and that the fields were close, meaning that the soil composition and the pedoclimatic conditions were the same. As such, any observed difference could only be caused by the different fertilization method.

As expected, one of the differences could be ascribed to the nitrogen metabolism. Indeed, the decrease found in organic sample of glutamine, asparagine and arginine, coupled with the increase of glutamate indicated a lowered activity of the nitrogen transport occurring from the roots toward the stem and the flowers of the plant [43,44]. Moreover, while it is not statistically significant to univariate analysis, fumarate covariate with conventional grown broccoli in PCA as shown in Figure 2. This is interesting since one biosynthetic pathway of arginine starts from glutamine and it includes the fumarate production. Since the correlation values of fumarate and arginine are rather close, this could be a further indication of the observed trend that brings to reduction change of nitrogen transport in organic grown broccoli samples. This hypothesis could be explained by the different degree of absorption of nitrogen from urea compared to manure, as reported in literature [45].

At the same time, the plant responded to this condition bolstering its defences, such as the content in glucosinolates, glucoraphanin and glucobrassicin, which were higher in organic grown plants. It is interesting to observe that not only the molecules were more abundant, but so were also many of their amino acidic precursors. As reported in the literature, the precursors of the core moiety are glucose, glycine, and methionine [46].

While methionine could not be quantified, both glucose and glycine content increased in organic samples, even if glucose increase was not statistically significant.

The observed variation in the amount of the aromatic amino acid levels, tyrosine and histidine which are precursors of several secondary metabolites belonging to the phenylpropanoid pathway [47,48], their increase in organic broccoli could be interpreted as a bolstering of this pathway. On the other hand, in the same samples phenylalanine, another precursor of phenylpropanoids, was present in a lower amount than in conventional plants as shown in Table 1. This apparent inconsistency could be explained by the fact that phenylalanine is the precursor of both aromatic amino acids as well as other secondary metabolites.

Another indication of the gearing of substrate reprogramming toward defence metabolites could be the increase of linolenic fatty acid. This molecule is the precursor of several volatile metabolites with defensive functions and Brassicaceae family contains several of them [49].

The carbohydrate metabolism was affected also by this adaptation: the increase of sucrose, which is usually coupled with a reduction of fructose levels, has been reported in literature to be in indication of the plant to store energy to be utilized for the production of molecules to be used against pathogens [50].

4. Conclusions

This study has shown that, in the absence of pathologies, there is a strong response of the metabolism of broccoli grown with organic practice, which stimulates the production of secondary metabolites, as observed by the increase in the concentration of glucosinolates, glucoraphanin, and glucobrassicin, as well as their precursor amino acids.

From the literature it is known that the nitrogen supply greatly influences the glucosinolate content. This is an important aspect for this study, as it allows to hypothesize that the slower release of nitrogen from manure compared to urea affects the plant growth which responds altering the metabolism of nitrogen transport.

Supplementary Materials: The following are available online at http://www.mdpi.com/2304-8158/9/7/945/s1, Figure S1: [1]H spectrum of *Brassica oleracea* var. *italica* hydroalcoholic extract, Figure S2: [1]H spectrum of *Brassica oleracea* var. *italica* [1]H spectrum of *Brassica oleracea* var. *italica* chloroform extract, Figure S3: [1]H-^{1}H TOCSY spectrum of *Brassica oleracea* var. *italica* hydroalcoholic extract, Figure S4: [1]H-^{1}H TOCSY spectrum of *Brassica oleracea* var. *italica* chloroform extract, Figure S5: ^{1}H-^{13}C HSQC spectrum of *Brassica oleracea* var. *italica* hydroalcoholic extract, Figure S6: ^{1}H-^{13}C HSQC spectrum of *Brassica oleracea* var. *italica* chloroform extract, Figure S7: ^{1}H-^{13}C HMBC spectrum of *Brassica oleracea* var. *italica* hydroalcoholic extract, Figure S8: ^{1}H-^{13}C HMBC spectrum of *Brassica oleracea* var. *italica* chloroform extract, Table S1: Metabolites identified in the ^{1}H NMR spectra of the *Brassica oleracea* var. *italica*.

Author Contributions: M.L., V.R. and F.S.: Conceptualization, V.R., F.M., M.L. and F.S.: Methodology, formal analysis, investigation, resources, A.D., M.L., G.L.-B., M.E.D.C. and F.S.: Data curation; M.L., F.S., A.S., M.E.D.C. and M.D.: Writing of the original manuscript; M.L., M.E.D.C., V.R., F.M., A.D., G.L.-B., A.S., M.D., and F.S.: Review and editing of the manuscript, project administration, M.L. and F.S.: Supervision. All authors made a substantial contribution to the work and approved its publication. All authors have read and agreed to the published version of the manuscript.

Funding: This research received no external funding.

Conflicts of Interest: The authors declare no conflict of interest.

References

1. World Health Organization. *Global Action Plan for the Prevention and Control of Noncommunicable Diseases 2013–2020*; World Health Organization: Geneva, Switzerland, 2013.
2. Avato, P.; Argentieri, M.P. Brassicaceae: A rich source of health improving phytochemicals. *Phytochem. Rev.* **2015**, *14*, 1019–1033. [CrossRef]
3. Santini, A.; Novellino, E. Nutraceuticals: Beyond the Diet Before the Drugs. *Curr. Bioact. Compd.* **2014**, *10*, 1–12. [CrossRef]
4. Santini, A.; Cammarata, S.M.; Capone, G.; Ianaro, A.; Tenore, G.C.; Pani, L.; Novellino, E. Nutraceuticals: Opening the debate for a regulatory framework. *Br. J. Clin. Pharmacol.* **2018**, *84*, 659–672. [CrossRef] [PubMed]
5. Santini, A.; Cicero, N. Development of Food Chemistry, Natural Products, and Nutrition Research: Targeting New Frontiers. *Foods* **2020**, *9*, 482. [CrossRef]
6. FAOSTAT. Food and Agriculture Organization of the United Nations, Data from 2014. 2017. Available online: http://www.fao.org/faostat/en/#data/QC (accessed on 30 May 2020).
7. Pennington, J.A.T.; Fisher, R.A. Food component profiles for fruit and vegetable subgroups. *J. Food Comp. Anal.* **2010**, *23*, 411–418. [CrossRef]
8. Cartea, M.E.; Francisco, M.; Soengas, P.; Velasco, P. Phenolic Compounds in Brassica Vegetables. *Molecules* **2011**, *16*, 251–280. [CrossRef]
9. Guzman, I.; Yousef, G.G.; Brown, A.F. Simultaneous extraction and quantitation of carotenoids, chlorophylls, and tocopherols in Brassica vegetables. *J. Agric. Food Chem.* **2012**, *60*, 7238–7244. [CrossRef]
10. Li, Z.; Lee, H.W.; Liang, X.; Liang, D.; Wang, Q.; Huang, D.; Ong, C.N. Profiling of Phenolic Compounds and Antioxidant Activity of 12 Cruciferous Vegetables. *Molecules* **2018**, *23*, 1139. [CrossRef]
11. Verkerk, R.; Schreiner, M.; Krumbein, A.; Ciska, E.; Holst, B.; Rowland, I.; De Schrijver, R.; Hansen, M.; Gerhaeuser, C.; Mithen, R.; et al. Glucosinolates in *Brassica* vegetables: The influence of the food supply chain on intake, bioavailability and human health. *Mol. Nutr. Food Res.* **2009**, *53*, S219. [CrossRef]
12. Clarke, D.B. Glucosinolates, structures and analysis in food. *Anal. Methods* **2010**, *2*, 310–325. [CrossRef]
13. Beck, T.K.; Jensen, S.; Bjoern, G.K.; Kidmose, U. The masking effect of sucrose on perception of bitter compounds in Brassica vegetables. *J. Sens. Stud.* **2014**, *29*, 190–200. [CrossRef]
14. Groenbaek, M.; Jensen, S.; Neugart, S.; Schreiner, M.; Kidmose, U.; Lakkenborg Kristensen, H. Influence of cultivar and fertilizer approach on curly kale (*Brassica oleracea* L. var. *sabellica*). Genetic diversity reflected in agronomic characteristics and phytochemical concentration. *J. Agric. Food Chem.* **2014**, *62*, 11393–11402.
15. Traka, M.; Mithen, R. Glucosinolates, isothiocyanates and human health. *Phytochem. Rev.* **2009**, *8*, 269–282. [CrossRef]
16. Veeranki, O.; Bhattacharya, A.; Tang, L.; Marshall, J.; Zhang, Y. Cruciferous vegetables, isothiocyanates, and prevention of bladder cancer. *Curr. Pharmacol. Rep.* **2015**, *1*, 272–282. [CrossRef] [PubMed]

17. Sánchez-Pujante, P.J.; Borja-Martínez, M.; Pedreño, M.Á.; Almagro, L. Biosynthesis and bioactivity of glucosinolates and their production in plant in vitro cultures. *Planta* **2017**, *246*, 19–32. [CrossRef] [PubMed]

18. Blažević, I.; Montaut, S.; Burčul, F.; Olsen, C.E.; Burow, M.; Rollin, P.; Agerbirk, N. Glucosinolate structural diversity, identification, chemical synthesis and metabolism in plants. *Phytochemistry* **2020**, *169*, 112100. [CrossRef]

19. Ramirez, D.; Abellán-Victorio, A.; Beretta, V.; Camargo, A.; Moreno, D.A. Functional Ingredients from *Brassicaceae* Species: Overview and Perspectives. *Int. J. Mol. Sci.* **2020**, *21*, 1998. [CrossRef] [PubMed]

20. Hanschen, F.S.; Klopsch, R.; Oliviero, T.; Schreiner, M.; Verkerk, R.; Dekker, M. Optimizing isothiocyanate formation during enzymatic glucosinolate breakdown by adjusting pH value, temperature and dilution in Brassica vegetables and Arabidopsis thaliana. *Sci. Rep.* **2017**, *7*, 40807. [CrossRef] [PubMed]

21. Raiola, A.; Errico, A.; Petruk, G.; Monti, D.M.; Barone, A.; Rigano, M.M. Bioactive Compounds in *Brassicaceae* Vegetables with a Role in the Prevention of Chronic Diseases. *Molecules* **2017**, *23*, 15. [CrossRef]

22. Prieto, M.A.; López, C.J.; Simal-Gandara, J. Glucosinolates: Molecular structure, breakdown, genetic, bioavailability, properties and healthy and adverse effects. *Adv. Food Nutr. Res.* **2019**, *90*, 305–350.

23. Quirante-Moya, S.; García-Ibañez, P.; Quirante-Moya, F.; Villaño, D.; Moreno, D.A. The Role of Brassica Bioactives on Human Health: Are We Studying It the Right Way? *Molecules* **2020**, *25*, 1591. [CrossRef] [PubMed]

24. Dinkova-Kostova, A.T.; Kostov, R.V. Glucosinolates and isothiocyanates in health and disease. *Trends Mol. Med.* **2012**, *18*, 337–347. [CrossRef] [PubMed]

25. Abbaoui, B.; Lucas, C.R.; Riedl, K.M.; Clinton, S.K.; Mortazavi, A. Cruciferous Vegetables, Isothiocyanates, and Bladder Cancer Prevention. *Mol. Nutr. Food Res.* **2018**, *62*, e1800079. [CrossRef] [PubMed]

26. Bayat Mokhtari, R.; Baluch, N.; Homayouni, T.S.; Morgatskaya, E.; Kumar, S.; Kazemi, P.; Yeger, H. The role of Sulforaphane in cancer chemoprevention and health benefits: A mini-review. *J. Cell Commun. Signal.* **2018**, *12*, 91–101. [CrossRef] [PubMed]

27. Soundararajan, P.; Kim, J.S. Anti-carcinogenic glucosinolates in cruciferous vegetables and their antagonistic effects on prevention of cancers. *Molecules* **2018**, *23*, 2983. [CrossRef]

28. Zhang, N.-Q.; Ho, S.C.; Mo, X.-F.; Lin, F.-Y.; Huang, W.-Q.; Luo, H.; Huang, J.; Zhang, C.-X. Glucosinolate and isothiocyanate intakes are inversely associated with breast cancer risk: A case–control study in China. *Br. J. Nutr.* **2018**, *119*, 957–964. [CrossRef]

29. Gründemann, C.; Huber, R. Chemoprevention with isothiocyanates—From bench to bedside. *Cancer Lett.* **2018**, *414*, 26–33. [CrossRef]

30. Traka, M.H.; Melchini, A.; Coode-Bate, J.; Al Kadhi, O.; Saha, S.; Defernez, M.; Troncoso-Rey, P.; Kibblewhite, H.; O'Neill, C.M.; Bernuzzi, F.; et al. Transcriptional changes in prostate of men on active surveillance after a 12-mo glucoraphanin-rich broccoli intervention-results from the Effect of Sulforaphane on prostate CAncer PrEvention (ESCAPE) randomized controlled trial. *Am. J. Clin. Nutr.* **2019**, *109*, 1133–1144. [CrossRef]

31. Burčul, F.; Generalić Mekinić, I.; Radan, M.; Rollin, P.; Blažević, I. Isothiocyanates: Cholinesterase inhibiting, antioxidant, and anti-inflammatory activity. *J. Enzym. Inhib. Med. Chem.* **2018**, *33*, 577–582. [CrossRef]

32. Oliviero, T.; Verkerk, R.; Dekker, M. Reply to "Dietary glucosinolates and risk of type 2 diabetes in 3 prospective cohort studies". *Am. J. Clin. Nutr.* **2018**, *108*, 425. [CrossRef]

33. Ma, L.; Liu, G.; Sampson, L.; Willett, W.C.; Hu, F.B.; Sun, Q. Dietary glucosinolates and risk of type 2 diabetes in 3 prospective cohort studies. *Am. J. Clin. Nutr.* **2018**, *107*, 617–625. [CrossRef] [PubMed]

34. Jones, R.B.; Imsic, M.; Franz, P.; Hale, G.; Tomkins, R.B. High nitrogen during growth reduced glucoraphanin and flavonol content in broccoli (*Brassica oleracea* var. *italica*) heads. *Austr. J. Exp. Agr.* **2007**, *47*, 1498–1505. [CrossRef]

35. Sciubba, F.; Di Cocco, M.E.; Gianferri, R.; Capuani, G.; De Salvador, F.R.; Fontanari, M.; Gorietti, D.; Delfini, M. Nuclear Magnetic Resonance-Based Metabolic Comparative Analysis of Two Apple Varieties with Different Resistances to Apple Scab Attacks. *J. Agric. Food Chem.* **2015**, *63*, 8339–8347. [CrossRef]

36. Sciubba, F.; Avanzato, D.; Vaccaro, A.; Capuani, G.; Spagnoli, M.; Di Cocco, M.E.; Tzareva, I.N.; Delfini, M. Monitoring of pistachio (*Pistacia vera*) ripening by high field nuclear magnetic resonance spectroscopy. *Nat. Prod. Res.* **2017**, *31*, 765–772. [CrossRef] [PubMed]

37. Sciubba, F.; Di Cocco, M.E.; Angori, G.; Spagnoli, M.; De Salvador, F.R.; Engel, P.; Delfini, M. NMR-based metabolic study of leaves of three species of *Actinidia* with different degrees of susceptibility to *Pseudomonas syringae* pv *actinidiae*. *Nat. Prod. Res.* **2019**. [CrossRef]

38. Lucarini, M.; Durazzo, A.; Sciubba, F.; Di Cocco, M.E.; Gianferri, R.; Alise, M.; Santini, A.; Delfini, M.; Lombardi Boccia, G. Stability of Meat protein type I collagen: Influence of pH, ionic strength and phenolic antioxitant. *Foods* **2020**, *4*, 480. [CrossRef]

39. Lucarini, M.; Sciubba, F.; Capitani, D.; Di Cocco, M.E.; D'Evoli, L.; Durazzo, A.; Delfini, M.; Lombardi Boccia, G. Role of catechin on collagen type I stability upon oxidation: A NMR approach. *Nat. Prod. Res.* **2020**, *34*, 53–62. [CrossRef]

40. Miccheli, A.; Ricciolini, R.; Piccolella, E.; Delfini, M.; Conti, F. Modulation of human lymphoblastoid B cell line by phorbol ester and sphingosine. A 31P-NMR study. *Biochim. Biophys. Acta. Mol. Cell Res.* **1991**, *1*, 29–35. [CrossRef]

41. Tomassini, A.; Sciubba, F.; Di Cocco, M.E.; Capuani, G.; Delfini, M.; Aureli, W.; Miccheli, A. 1H NMR-Based Metabolomics Reveals a Pedoclimatic Metabolic Imprinting in Ready-to-Drink Carrot Juices. *J. Agric. Food Chem.* **2016**, *64*, 5284–5291. [CrossRef]

42. Wishart, D.S.; Tzur, D.; Knox, C.; Eisner, R.; Guo, A.C.; Young, N.; Cheng, D.; Jewell, K.; Arndt, D.; Sawhney, S.; et al. HMDB: The Human Metabolome Database. *Nucleic Acids Res.* **2007**, *35*, D521–D526. [CrossRef]

43. Joi, K.W. Ammonia, glutamine and asparagine a carbon-nitrogen interface. *Can. J. Bot.* **1988**, *6*, 2103–2109. [CrossRef]

44. Krapp, A. Plant nitrogen assimilation and its regulation: A complex puzzle with missing pieces. *Curr. Opin. Plant Biol.* **2015**, *25*, 115–122. [CrossRef] [PubMed]

45. Irshad, M.; Yamamoto, S.; Eneji, A.E.; Endo, T.; Honna, T. Urea and manure effect on growth and mineral contents of maize under saline conditions. *J. Plant Nutr.* **2002**, *25*, 189–200. [CrossRef]

46. Ishida, M.; Hara, M.; Fukino, N.; Kakizaki, T.; Morimitsu, Y. Glucosinolate metabolism, functionality and breeding for the improvement of Brassicaceae vegetables. *Breed. Sci.* **2014**, *64*, 48–59. [CrossRef] [PubMed]

47. Michalek, S.; Klebel, C.; Treutter, D. Stimulation of phenylpropanoid biosynthesis in apple (*Malus domestica* borkh.) by abiotic elicitors. *Eur. J. Hortic. Sci.* **2005**, *70*, 116–120.

48. Qudsia, K.; Ishtiaq, H.; Hamid, L.S.; Arshad, J. Antifungal activity of flavonoids isolated from mango (*Mangifera indica* L.) leaves. *Nat. Prod. Res.* **2010**, *24*, 1907–1914.

49. Kessler, A.; Baldwin, I.T. Defensive function of herbivore-induced plant volatile emissions in nature. *Science* **2001**, *291*, 2141–2144. [CrossRef]

50. Kanwar, P.; Jha, G. Alterations in plant sugar metabolism: Signatory of pathogen attack. *Planta* **2019**, *249*, 305–318. [CrossRef]

 *foods*

Article

Three Types of Beetroot Products Enriched with Lactic Acid Bacteria

Vasilica Barbu [1], **Mihaela Cotârleț** [1], **Carmen Alina Bolea** [1], **Alina Cantaragiu** [2], **Doina Georgeta Andronoiu** [1], **Gabriela Elena Bahrim** [1] and **Elena Enachi** [1,*]

[1] Faculty of Food Science and Engineering, Dunărea de Jos University of Galati, 111 Domneasca Street, 800201 Galati, Romania; vasilica.barbu@ugal.ro (V.B.); mihaela.cotarlet@ugal.ro (M.C.); carmen.bolea@ugal.ro (C.A.B.); georgeta.andronoiu@ugal.ro (D.G.A.); gabriela.bahrim@ugal.ro (G.E.B.)

[2] Research and Development Center for Thermoset Matrix Composites, Dunărea de Jos University of Galati, 111 Domneasca Street, 800201 Galati, Romania; alina.cantaragiu@ugal.ro

* Correspondence: elena.ionita@ugal.ro; Tel.: +40-336-130-182

Received: 19 May 2020; Accepted: 11 June 2020; Published: 14 June 2020

Abstract: Beetroot (*Beta vulgaris* L.) represents a very rich source of bioactive compounds such as phenolic compounds and carotenoids, among which the most important being betalains, mainly betacyanins and betaxanthins. The beetroot matrix was used in a fresh or dried form or as lyophilized powder. A 10^{12} CFU/g inoculum of *Lactobacillus plantarum* MIUG BL3 culture was sprayed on the vegetal tissue. The lactic acid bacteria (LAB) viability for all the products was evaluated over 21 days, by microbiological culture methods. The antioxidant activity of the obtained food products was correlated to the betalains content and the viability of LAB. The content of polyphenolic compounds varied between 225.7 and 1314.7 mg L^{-1}, hence revealing a high content of bioactive compounds. Through the confocal laser scanning microscopy analysis, a large number of viable probiotic cells were observed in all the variants but especially in the fresh red beet cubes. After 21 days of refrigeration, the high content of *Lb. plantarum* (CFU per gram) of the food products was attributed to the biocompounds and the nutrients of the vegetal matrix that somehow protected the bacterial cells, and thus maintained their viability. The obtained food products enriched with probiotic LAB can be regarded as new functional food products due to the beneficial properties they possessed throughout the undertaken experiments.

Keywords: beetroot; lactic acid bacteria; betalains; functional food products

---

1. Introduction

Beetroot (or red beet) is an edible taproot of *Beta vulgaris* L. subsp. *vulgaris* (var. *conditiva*) species that is phylogenetic framed in the *Betoideae* subfamily, in the *Amaranthaceae* family and is the most important crop of the large order *Caryophyllales*. The specie has many varieties that are grown throughout the Americas, Europe, and Asia. Beetroot, through its rich content in betalains, phenolic compounds, and carotenoids, is an important source of biocompounds with antioxidant properties and may represent a strong alternative to synthetic dyes. Natural pigments without any toxic effect are extremely valuable and they have important applications in the field of medicine, food, cosmetic or pharmaceutical industries. Betalains are a group of secondary plant phenolic metabolites, being water-soluble vacuolar chromoalkaloids found in many plants with health benefits for humans, especially regarding their antioxidant, anti-inflammatory, antiviral, even anti-tumoral activities [1–3]. Tesoriere et al. [4] and Reddy et al. [5] also proved the betalains role concerning the inhibition of lipid peroxidation, a very important process in the cholesterol metabolism. The main betalains that are found in beetroot are the red-violet betacyanins and yellow-orange betaxanthins [6,7] and, so far, several other natural derivatives have been described [8–10]. Among betacyanins, betanin, known as CI Natural Red 33 or E162, is the only betalain approved to be used in food, being almost totally extracted from beetroot crops [11] whereas among betaxanthins, indicaxanthin is the most important. Apart from betalains, other types of phenolic compounds have been identified such as small amounts of gallic, syringic, and caffeic acids and flavonoids [12]. Therefore, beetroot extract has shown a strong phytochemicals pattern which resulted in an increased researchers' interest in the last decades. For these reasons, red beet represents a rich source of bioactive compounds and could be used to develop functional foods [1,6].

Lactic acid bacteria (LAB), especially Lactobacilli and their metabolites play a key role in improving the microbiological quality and also in increasing the shelf life period of many functional, fermented food products. Lately, the biotechnological studies in the food industry domain have designed several commercial products that contained a single probiotic starter strain or a bacterial consortium. Furthermore, during recent years, it was thoroughly proved that the regular consumption of viable probiotics can determine a number of health benefits such as: a decreasing cholesterol level [13,14], anti-diabetic effect [15–17], alleviating diarrhea and constipation, improving the lactose intolerance and the human gut microbiome, in general [18,19], strengthening the immune system [20] or even antitumoral effects [21,22].

This study followed the design of some new functional food products by combining the red beetroot under different forms with LAB, with a strong focus on the beneficial effects for health. The main objectives of the present research were to analyze these original products from the microbiological, biochemical, textural, ultrastructural point of view in order to highlight their functionality. The CIE color parameters as well as digestibility were also studied in order to further highlight the nutritional value of the newly designed products.

2. Materials and Methods

2.1. Lactic Acid Bacteria Strain

The *Lactobacillus plantarum* BL3 strain was used for this analysis from the Microorganisms Collection of the Bioaliment Research Platform (acronym MIUG) at −70 °C in MRS broth (De Mann, Rogosa and Sharpe—Sigma Aldrich, Darmstadt, Germany) supplemented with 20% (*v/v*) glycerol. The primary culture of lactic acid bacteria (LAB) was obtained after its cultivation on MRS broth (for 12 h at 37 °C) in order to obtain the mid-logarithmic phase cultures that were further used in our experiments. The cells were harvested by centrifugation at 4800 rpm min^{-1}, at 4 °C, for 10 min and were washed twice with sterile 0.85% saline solution. A 10^{12} CFU mL^{-1} inoculum of *Lactobacillus plantarum* MIUG BL3 was used to spray the vegetal tissue.

2.2. Sample Preparation

The beetroot that was used for all the experiments was purchased from the local market. The beetroot was washed, peeled, and cut under aseptic conditions by using a food processor into cubes (with an edge of 5 mm ± 0.2 length), as slices or as thin chips (with an area of 2–3 cm^2). The vegetal matrix was split into equal amounts of 100 g each, transferred to sterile Petri dishes and exposed at the UV light, for 30 min, by using a SafeFastElite 215S Microbiological Safety Cabinet (Faster, Cornaredo, Italy). On the fresh beetroot cubes, the bacterial suspension was sprayed in a 5:1 (*w/v*) ratio (F variant). After that, half of these samples (F) were packed into sterile, single use plastic zip-lock bags and refrigerated at 4 °C for the microbiological analysis in order to determine the shelf life. The other half of samples were frozen at −80 °C (Angelantoni Platinum 500+, ALS, Massa Martana, Italy) and then freeze-dried at 10 mBar and −50 °C, for 48 h, using a CHRIST ALPHA 1-4 LD plus equipment (Martin Christ, GmbH, Osterode am Harz, Germany) until constant weight was obtained (with a 4.25 ± 0.2% water content). The freeze-dried samples were then ground, portioned, and packed identically in quantities of 10 g powder (FDP) per sachet and were kept at the room temperature.

The chips were slowly dried (at 42 °C for 12 h) until a 12.55 ± 0.2% moisture content was reached, by using a cabinet Stericell 111 air dryer (Medcenter GmbH, München, Germany). On the dried chips, the LAB suspension (*Lactobacillus plantarum* BL3) was sprayed in a ratio of 3:1 (*w/v*), hence obtaining the dried chips (DC) variant. These samples (DC) were transferred into sterile, single use plastic zip-lock bags that were kept at room temperature in order to perform the microbiological analysis.

Thus, all three variants: fresh (F), dried chips (DC), and freeze-dried powder (FDP) have LAB in their contents. In the DC variant, LAB were sprayed after drying the chips, in the FDP variant, LAB were sprayed before lyophilization and grinding, and in the fresh variant (F), LAB were sprayed on the beetroot cubes and kept in this form at 4 °C.

2.3. Lactic Acid Bacteria (LAB) Viability

In order to estimate the shelf life period, the viability of the LAB in all the variants was weekly evaluated, by cultural methods, for a period of 21 days. The samples were homogenized in a 0.85% sterile saline solution using a Pulsifier equipment (Microgen Bioproduct, London, UK) at medium speed and maintained for 5 min. The homogenized samples were serially diluted with 0.85% sterile saline solution and spread over the MRS agar supplemented with 2% $CaCO_3$. The plates were incubated at 37 °C for 48 h and the colonies were counted. The experiment was conducted in triplicate. The viability of *Lb. plantarum* BL3 was expressed as the log_{10} of the mean number of the colony forming units (that were counted for each dilution in three different plates) (CFU g^{-1}).

2.4. Scanning Electron Microscopy (SEM)

The samples' ultrastructures enriched with LAB were analyzed by scanning electron microscopy. The samples were attached on the aluminum stubs with double adhesive carbon conductive tape (12 mm W × 5 mL) and gold coated with 5 nm as thickness in an argon atmosphere by using SPI Supplies (USA) sputter coater. The surface micrographs of all the samples were obtained using the FEI Quanta 200 SEM (Fei Europe B.v. Eindhoven, The Netherlands) with a plasma current intensity of 18 mA, a pressure of 6 mBar, and a spot size of 10mm as working distance. The SEM images were taken at different magnifications between 100× and 5000×.

2.5. Confocal Laser Scanning Microscopy

The confocal images of the functional food products based on beetroot and LAB, were acquired with a Zeiss confocal laser scanning system (LSM 710) equipped with a diode laser (405 nm), Ar-laser (458, 488, 514 nm), DPSS laser (diode pumped solid state—561 nm) and HeNe-laser (633 nm). In order to observe in detail the vegetal microstructures and the *Lb. plantarum* BL3 cells, the Live/Dead Backlight bacterial viability stain kit (Molecular Probes, Eugene, OR, USA) was used according to the manufacturer's instructions so that one drop was applied directly to the surface of each sample. It consisted of a two nucleic acid-binding stains mixture: SYTO9 which stained all the viable bacteria (shown in green), while the propidium iodide stained the non-viable bacteria (shown in red), after 15 min of dark incubation [23]. The excitation and emission wavelengths were 480 and 500 nm for SYTO9 and 490 and 635 nm for propidium iodide, respectively. The samples were observed with a Zeiss Axio Observer Z1 inverted microscope equipped with a 40× apochromat objective (numerical aperture 1.4). The 3D images were rendered and analyzed by a ZEN 2012 SP1 Black edition software. A minimum of twenty fields were evaluated, all the viability counts being determined in two independent experiments, with each assay being performed in triplicate.

2.6. Hardness Measurement

The texture of the beetroot dried chips (including hardness, porosity, crispness, chewiness, springiness) was determined using a Brookfield CT3-1000 texture analyzer (Ametek Brookfield, Middleborough, MA, USA), equipped with a 4 mm diameter steel cylinder (TA44). The stress at the maximum force was evaluated using a puncture test in the center of each dried chips. The samples had

30 mm length, 20 mm width, and 3 mm depth. The Texture Profile Analysis method (TPA) was used to determine the hardness and springiness of the samples. The stress at the maximum force was related to the hardness of the beetroot chips. The test speed was set at 1 mm s^{-1}, with a target distance of 2 mm and a trigger load of 0.067 N. Before each experiment, a randomly chosen single beetroot chip was placed on the bottom parallel plate and compressed. The compression experiments were performed in 15 replications. The average force and energy required to cause deformation were determined on the basis of force-deformation curves. The beetroot chips were analyzed weekly, for a period of 21 days.

2.7. Betalains Quantification

In order to extract the pigments from the plant tissue, a weak acid solution (0.5% citric acid and 0.1% ascorbic acid) was used in accordance to Neagu and Barbu [24] so that the maximum content was obtained with the minimum pigment degradation. The liquid/solid ratio used in this study was 5:1 (*w/v*) and the mixture was homogenized for 30 min by magnetic stirring and afterwards filtered. The operation was repeated three times and the total extract was evaporated under mild conditions under vacuum at a temperature that did not exceed 40 °C with a Christ RVC 2-18 equipment (Martin Christ, Germany) until the aqueous extract was concentrated. The betalains content (B) of the extracts were determined spectrophotometrically at a wavelength corresponding to the maximum absorption of each of the betalains with a JENWAY 6505 UV/Vis equipment, according to the Lambert Beer's law: $A = \log(I_{10}/I) = \varepsilon \times L \times c$. The following formula was further used: $B\ (mg/L) = [(A_i \times F_d \times MW \times 1000)/(\varepsilon \times l)]$, where A_i is the absorption at 538 for betacyanins and 480 nm for betaxanthins, F_d is the dilution factor and l the path length of the cuvette (1 cm). To quantify the betacyanins (BC) and betaxanthins (BX), the molecular weights (MW) and molar extinction coefficients (ε) of the representative compounds were undertaken i.e., betacyanins (ε = 60,000 L mol cm^{-1} in H_2O; MW = 550 g/mol) and betaxanthins (ε = 48,000 L/mol cm in H_2O; MW = 339 g/mol). The betalains content was assessed with the formula (B) (mg/L) = BC + BX [6,25,26]. Fresh red beet cubes without LAB were analyzed as the control sample.

2.8. Color Measurement

In order to assess the color difference between the samples and also the Hunter color L*, a*, and b* parameters a CR-410 Chroma Meter (Konica Minolta Sensing Americas, Ramsey, NJ, USA) was used. The values were determined after the samples were ground into a fine powder. The information given by the L*, a* and b* parameters was generally expressed as the total color of the beetroot samples, the positive L* values representing the brightness and negative values representing lusterless or dullness, a* for the redness (+) to greenness (−), and b* for the yellowness (+) to blueness (−). The second method for the browning assessment was assessed by using the following formula: Browning Index (BI) = (100 (x − 0.31))/0.17, where x = (a* + 1.75L*)/(5.645L* + a* − 0.3012b*), according to Chandran et al. [27].

2.9. Antioxidant Activity

The antioxidant capacity was determined as described by Yuan et al. [28], by using a DPPH (2,2-diphenyl-1-picrylhydrazyl, Fluka Chemie, Fluka Chemie GmbH, Buchs, Switzerland) methanolic solution (0.1 M) with 30 min as the reaction time, and quantified using a Trolox (6-hydroxy-2,5,7,8-tetramethylchroman-2-carboxylic acid) calibration curve, under the same conditions. The antioxidant activity of the beetroot extract was measured by a spectrophotometric method at 517 nm. The percent inhibition of DPPH was calculated as follows: Antioxidant activity (%) = ((Absorbance of the blank−Absorbance of the sample)/Absorbance of blank) × 100. Fresh red beet cubes without LAB were analyzed as the control sample.

2.10. Total Phenolic Content

The total phenolic content (TPC) was determined using the Folin-Ciocalteu reagent and gallic acid as a standard after the method of Turturică et al. [29]. The beetroot extracts were totally dissolved in ddH$_2$O (1 mg/mL). Shortly after, 200 µL aliquots of the resulting solution were mixed with 125 µL of 2 N Folin–Ciocalteau's phenol reagent, diluted 1:2 (*v/v*). After 3 min of mixing, 125 µL of 20% Na$_2$CO$_3$ and 550 µL of deionized water were added. The resulting mixture was kept for 30 min in the dark, at room temperature; after that the mixture was centrifuged at 8200× *g* for 10 min. The absorbance was measured at 765 nm. The results were expressed as mg of gallic acid equivalents (GAE) per liter. Fresh red beet cubes without LAB were analyzed as the control sample. Each sample was measured in triplicate.

2.11. Digestibility

To achieve the in vitro digestibility of the obtained products, a static method was used as stated by Croitoru et al. [30], with small modifications. The products were shredded and afterwards mixed with a 10 mM Tris-HCl, pH 7.7, at a ratio of 1g of product to 10 mL of buffer solution. In order to thoroughly simulate the digestive conditions, a gastric mixture containing 20 mg of pork pepsin and 20 mL of HCl 0.1 N to reach the pH 2.0. The samples were further incubated at 37 °C on an orbital shaker (Optic Ivymen System, Grupo-Selecta, Barcelona, Spain) at 170 rpm. Regarding the intestinal digestibility, the mixture contained 40 mg of pancreatic enzymes and 20 mL of sodium bicarbonate 1M, the mixture having a pH of 7.7. The determination of betalains content was assessed as described previously.

2.12. Statistical Analysis of Data

Unless otherwise stated, the data reported in this study represent the averages of triplicate analysis and were reported as mean ± standard deviation. The analysis of variance (ANOVA) ($p < 0.05$) was carried out to assess the significant differences between values.

3. Results and Discussion

3.1. LAB Viability

Tripathi and Giri [31] stated that many factors were found to influence the viability of probiotic microorganisms in food products during production, processing, and storage such as fermentation and storage conditions (temperature, pH, medium, oxygen, etc.) protective agents, microencapsulation methods, food ingredients, processing operations (drying, freeze-drying etc.) [31].

The comparison survival of LAB between all the obtained products indicated a different viability behavior which in term depended on the processing method. The initial LAB content in all the experimental variants was evaluated immediately after the sample's preparation. The content presented values between 5.14×10^7 CFU g^{-1} (in fresh cubes of beetroot) and 7.05×10^7 CFU g^{-1} (in dried beetroot chips). The higher value in the DC variant was at this rate, due to the fact that LAB suspension was sprayed, in a ratio of 3:1 (*w/v*), after the drying process. Although, at the beginning of the experiment, the LAB content started from a larger value in the DC version, the microbiological analysis showed that during storage, at room temperature, the number of probiotic lactic bacteria was slightly reduced, although the content was rather constant being still at a large magnitude order (10^7 CFU g^{-1}) even after three weeks (Figure 1). Surprisingly, the evolution of the *Lb. plantarum* BL3 strain in the fresh variant, registered a 100-fold increase from 5.14×10^7 CFU g^{-1} to 1.46×10^9 CFU g^{-1} ($p \leq 0.05$), during the storage period at 4 °C, value attributed to the biochemical environmental conditions that were favorable for development and multiplication. Similar results were obtained also in the case of a patented product based on carrot slices fermented with *Lactobacillus strain* NCIMB 40,450 in which, after 28 days of fermentation, a concentration of $1–6 \times 10^8$ CFU g^{-1} was reached [32] (EU patent no: 0536851). Regarding our results, in the fresh vegetal matrix, the tissue nutrients (carbohydrates and proteins as a source of C and N, respectively) are more accessible for the lactic acid bacteria, which explains the increase of the LAB number in the F variant. Because the accessibility of the nutrients was higher, the lactic fermentation process took place, which contributed to the special organoleptic impression of this functional product variant. The fresh product had a sourer taste and a softer texture compared to the other samples. It should be also emphasized that the lactic microbiota has not dropped below 10^6 CFU g^{-1} in any of the proposed functional food products, even after three weeks, which allowed us to appreciate the shelf life of these products. The maintenance of a sufficiently large number of CFU per gram of product in the DC and FDP variants could be explained by the fact that the biocompounds and the nutrients of the vegetal matrix microencapsulated and protected the bacterial cells, and thus maintained their viability. The freeze-dried supplementary products better preserved the morphology of the probiotic microcapsules [31]. Additionally, the presence of yeasts, molds, or other mesophilic aerobic bacteria throughout the entire storage period has not been reported, probably because the lactic acid bacteria had an antimicrobial effect that inhibited the alteration microbiota, making these valuable products safe for consumption.

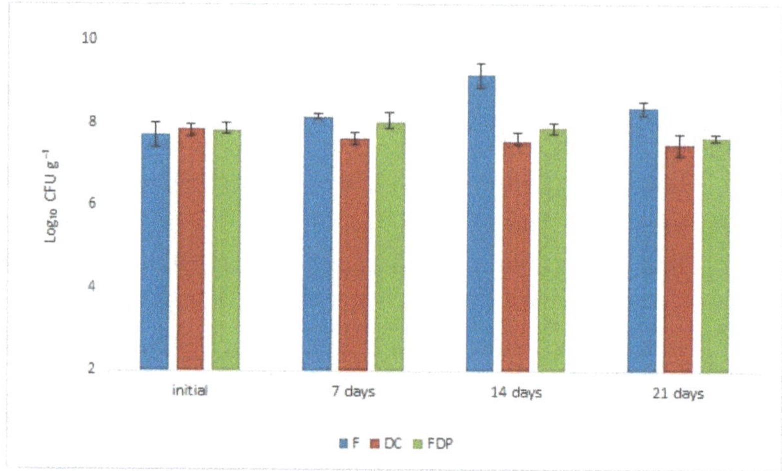

Figure 1. Viability of *Lactobacillus plantarum* BL3 in the experimental variants during 21 days of storage: F—Fresh beetroot cubes, DC—dried chips of red beet, FDP—freeze dried powder (data are the means ± SD).

3.2. Scanning Electron Microscopy (SEM)

The scanning electron microscopy images showed the vegetal tissue with its isodiametric parenchymal cells and their whole cell walls (in fresh beetroot cubes—Figure 2A,a). The drying process at mild temperatures (42 °C) resulted in the changing of the cells shape, loss of turgor (Figure 2B,a), these modifications being more evidenced in the freeze-dried powder (Figure 2C,a) where the cells were smaller, at the same magnification (200×).

Much more interesting than the microscopic appearance of the plant tissue subjected to mild processing techniques was the appearance of the *Lb. plantarum* BL3 bacterial biofilm that adhered to the large surfaces of the plant cells walls (Figure 2b,c). The main characteristic of the biofilms was the formation of an extracellular polysaccharide (EPS) matrix, which provided protection of biocompounds and helped to create a microenvironment for the metabolic interaction of the population [33,34]. These extracellular capsules could be clearly observed for all the experimental variants (Figure 2A–C), especially at 5000× magnification (Figure 2c).

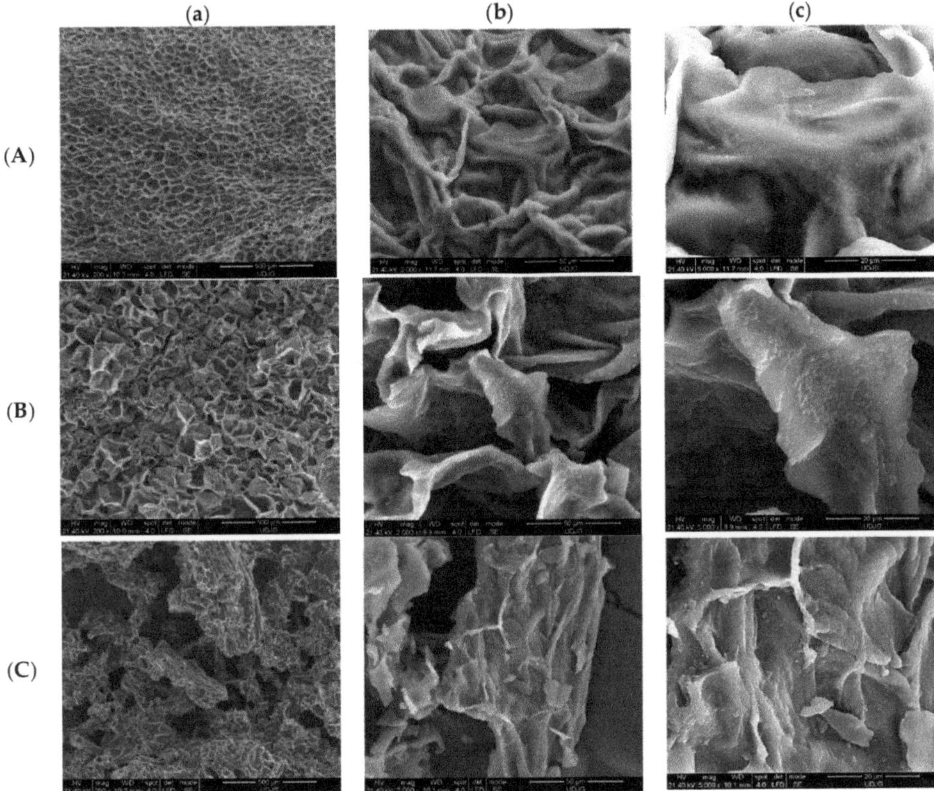

Figure 2. SEM images of the samples' ultrastructure: (**A**)—fresh cubes with lactic acid bacteria (LAB); (**B**)—dried chips with LAB; (**C**)—freeze-dried powder with LAB; (**a**)—200×, (**b**)—2000×, (**c**)—5000×.

The biofilm was at its third stage after seven days after the inoculation and was characterized by the aggregation of cells into microcolonies that resulted from the simultaneous division and growth of the microorganisms. EPS helped to strengthen the bond between the bacteria and the substratum and stabilized the colony against any environmental stress. Thus, it could be observed that the amount of exopolysaccharides was even greater and the microcolonies were becoming more extensive as the processing technology was stronger (Figure 2c comparing between A, B, C).

3.3. Confocal Laser Scanning Microscopy

Confocal laser scanning microscopy images showed a huge number of viable LAB (colored in light-green) in all the variants but especially in the fresh red beet cubes (Figure 3A), where a significant multiplication of *Lb. plantarum* BL3 cells were observed during the refrigeration period, which was probably due to the rich environment offered by the beetroot sap. Moreover, the high moisture, high content of minerals, vitamins, biocompounds, and carbohydrates provided a supportive environment for the multiplication of *Lb plantarum* BL3 bacteria. About 80–90% of the *Lb plantarum* BL3 cells were viable in the dried chips (B) or in the freeze-dried powder (C), being observable by using the fluorescent dyes included in the LIVE/DEAD BacLight kit. These results were well correlated to those obtained by Moreno et al. [23] who assessed the LAB quantitative evaluation by indirect cultural methods. Additionally, our results are in accordance with the US FDA recommendations that state that the minimum probiotic count in a probiotic food should be at least 10^6 CFU mL^{-1}. Depending on the ingested amount (100 g in F and DC variants and 10 g in FDP supplement) and by also taking into account the effect of storage on the probiotic viability, a daily intake of 10^8–10^9 probiotic microorganisms will ensure the achievement of a probiotic action upon the human organism [31].

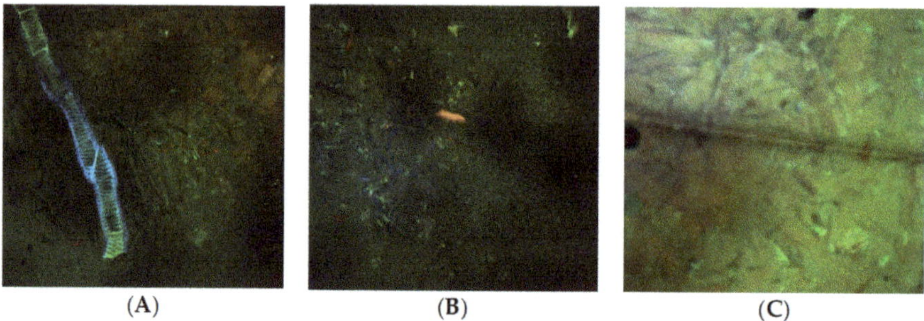

(**A**)   (**B**)   (**C**)

Figure 3. CLSM images of samples microstructure: (**A**)—fresh cubes with LAB; (**B**)—dried chips with LAB; (**C**)—freeze dried powder with LAB (*Lb. plantarum* BL3 viable cells—in green and the non-viable bacteria in red).

3.4. Measurement of Hardness

The results of TPA are presented in Table 1. The data represented the mean of three determinations. The hardness registered a slight increase during storage, from 0.73 N to 0.77 N, maybe due to the water loss. However, these values indicated a soft texture of chips, explained by the fact that during the dehydration process the equilibrum humidity was not achieved. At the same time, springiness declined from 1.57 mm to 1.52 mm, so that during storage the samples lost their elasticity. The texture parameters of beetroot chips did not vary significantly, asuring a constant texture quality during storage. Other characteristics correlated with the springiness were: gumminess index: 0.51 N, chewiness index: 0.41 N, similar to those obtained by Cui et al. [35] on apple chips.

Table 1. Texture Profile Analysis of beetroot dried chips enriched with *Lb. plantarum* BL3.

Texture Parameter	7 Days	14 Days	21 Days
Hardness, N	0.73 ± 0.008	0.75 ± 0.008	0.77 ± 0.008
Springiness, mm	1.57 ± 0.005	1.56 ± 0.008	1.52 ± 0.005

3.5. Betalains Quantification, Total Polyphenols Content, and Antioxidant Activity

The content of betacyanin (BC) and betaxanthin (BX) in the fresh beetroot cubes without LAB (control samples) was 147.90 ± 5.204 mg L^{-1} and 64.68 ± 1.004 mg L^{-1}, respectively (Table 2). The betacyanins represented 69.5% of the total betalalains. The DC and the FDP variant registered significantly higher BC and BX content compared to the F variant, respectively. For the DC variant, the BC and the BX contents were 349.25 ± 1.082 mg L^{-1} and 298.38 ± 5.854 mg L^{-1} whereas the FDP variant had a BC content of 689.79 ± 4.321 mg L^{-1} and a BX content of 786.69 ± 5.625 mg L^{-1}. The addition of the selected LAB with probiotic effect caused a significant increase of the betalains content probably due to the increased extractability of these pigments. Betalains highest content was recorded in the FDP samples, the increase being more obvious in the betaxanthin case, and as such in the FDP variant the proportion of the two betalains classes was inverted (Figure 4). It is known that betaxanthins are more stable than betacyanins under the temperature and freeze-drying conditions [27].

Table 2. Phytochemical features of the F, DC, and FDP experimental variants—The experimental variants were compared to the control (fresh beetroot cubes without LAB).

Variants	BC (mg L^{-1})	BX (mg L^{-1})	TPC (mg Gallic Acid L^{-1})	Antioxidant Activity (%)
Control	147.90 ± 5.204	64.68 ± 1.004	225.70 ± 0.034	20.19
F	226.18 ± 2.002	185.12 ± 1.229	418.30 ± 0.045	22.13
DC	349.25 ± 1.082	298.38 ± 5.854	635.80 ± 0.005	33.51
FDP	689.79 ± 4.321	786.69 ± 5.625	1314.70 ± 0.025	56.85

Note: F—fresh beetroot cubes; DC—dried beetroot chips; FDP—freeze-dried beetroot powder; BC—betacyanin, BX—betaxanthin, TPC—total polyphenols content.

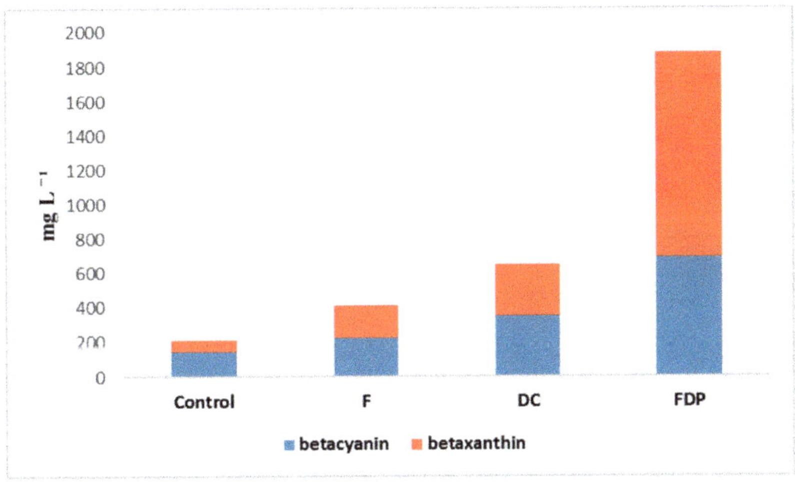

Figure 4. The betacyanin (BC) and betaxanthin (BX) content and their ratio in the experimental variants.

The betalains content varies very much depending on the harvesting time, climatic conditions, variety, plant parts, maturation stage and extraction, or preservation methods [25,36]. Bucur et al. [25] showed that spring varieties had a much higher betalains content (almost double) compared to autumn red beet varieties due to the high temperature of the soil during the summer which in term caused the degradation by isomerization, decarboxylation, or by hydrolysis of the betalain molecule to betalamic acid [25,37,38]. The temperature and pH of the extraction medium and solid: liquid ratio are also very important factors that may modify the betalains values [24,36,39]. Freezing preservation has also been shown to halve the content of betalains in red beets [25]. Our results were similar to those obtained by Slavov et al. [40] when the whole beetroot was used as substrate.

When it comes to beetroot, the antioxidant activity was usually correlated with the betalains content [25]. Moreover, it is known that processed fruits and vegetables have lower antioxidant activity compared to the raw or fresh due to vitamin C degradation during processing [38]. Extremely variable values of antioxidant activity were also obtained by Kushwaha et al. [36]. The parameter depends on many factors related to variety or extraction methods, but the highest antioxidant (86.34%) activity was obtained at 50 °C and pH 3.5 [36]. Our functional food products registered a slight increase of the antioxidant activity compared to control samples, with the FDP displaying the highest values, 56.85% (Table 2).

The content of polyphenolic compounds varied from 225.7 to 1214.7 mg L^{-1} (Table 2), which was rather similar compared to the findings of Wruss et al. [7]. Kushwaha et al. [36] observed that with the increase of the extraction temperature from 40 to 55 °C, a slight increase of the phenolic extraction rate was also observed which might be due to the softening of tissue responsible for the weakening of the phenol–protein and phenol polysaccharide interaction in the plant tissue. The drying temperature of 42 °C favored the development of the lactic microbiota as well as the extraction of betalains and polyphenols.

3.6. Measurement of Color

The food quality stability requires a better understanding of the parameters that may influence it. The color plays an important role in the visual recognition and assessment of the surfaces and it has a great influence on the appearance and acceptance of food products. The beet root powder enriched with LAB, besides the intake of probiotics, could be used to improve the red color of dressings, mousses, jams, soups, jellies, ice creams, sweets, or breakfast cereals [27,41]. There are different modalities, arrangements, or formulas to appreciate the color of red beet by using L, a, and b CIE color parameters. Hue value was calculated as the angle that had the b/a tangent, such as the values from 0° and 90° corresponded to the hue that varied from pure red to pure yellow while the hue values between 90° and 180° corresponded to the hue that varied from pure yellow to pure green. From this perspective, the lowest hue value was determined in the case of fresh red beet cubes enriched with LAB (Table 3). In this variant, the red color was the most stabile probably because of the lactic fermentation produced by the *Lb. plantarum* BL3. This also correlated with the increased value of parameter a (Table 3). The fermentation in the fresh cubes variant (F) determined an increased lightness and red chromaticity, when compared to dried chips. The long drying, even at gentle temperatures, clearly affected the color of the functional product in our study. As mentioned by other researchers [27], the Hunter a/b ratio was found to be the best parameter to express the color degradation as quantitatively and correctly as possible [27]. Moreover, these authors showed that the color degradation measured as Hunter 'a/b' value followed a first order kinetics, where the rate constant increased with the increasing of temperature [27]. The a/b ratio of the chips obtained after 12 h of drying at 42 °C was 2.7 (Table 3), a result that was similar to those obtained by Chandran et al. [27] after 60 min of drying at 90 °C. It was also reported that the yellow pigments of beet root, betaxanthins, were more stable than the betacyanins (red pigments), although the degradation of both pigments was proportional [27,42]. The FDP variant showed intermediate values of all the color parameters compared to the F and DC variants, however closer to those of the fresh variant. Thus, it can be stated that drying, even at mild temperatures (42 °C for 12 h), affects more the CIE color parameters compared to the freeze-drying method. As a result, the FDP powder proved to possess ideal properties to be regarded as a food additive.

Table 3. The drying method effect on the color of probiotic enriched beetroot samples.

	L	a	b	x	BI	a/b	b/a	Hue Value (tg⁻)
F	35.64 ± 0.34	62.76 ± 0.23	19.58 ± 0.40	0.485 ± 0.18	114.117 ± 0.32	3.20	0.3119	17.32°
DC	32.83 ± 0.18	45.25 ± 0.50	16.51 ± 0.40	0.455 ± 0.26	85.58 ± 0.31	2.70	0.3907	21.34°
FDP	34.63 ± 0.32	57.61 ± 0.49	18.73 ± 0.10	0.477 ± 0.17	98.235 ± 0.21	3.07	0.3251	18.00°

3.7. Digestibility of the Beetroot Products

The in vitro digestibility of the beetroot products has sought to assess the betalains behavior in both the gastric and intestinal juices. Prior to the absorption, the betalains could be hydrolysed in the gastrointestinal tract due to the environmental impact or due to some enzymes that are able to hydrolyze the betacyanins and betaxanthins. Regarding the other phytochemicals, previous studies have indicated that the conditions in the digestive tract (temperature, pH) and the β-glucosidases of the microflora could cause the hydrolysis of several important bonds. Another important factor may be the activity of the cytosolic β-glycosidases in the intestinal mucosal cell on the phytochemicals absorbed by enterocytes [43]. Betacyanins glycosides may be transferred to the gastrointestinal tract mucosal cells and subsequently hydrolyzed by cytosolic enzymes. It is also possible that the betalains aglycans reach the large intestine where they can be hydrolysed by β-glucosidases produced by the bacteria present in the colon but also by the alkaline environment [44]. The biologically active compounds in vitro digestibility in the gastrointestinal tract (Figure 5a–d) is a complex process, with a major impact on their release, distribution and bioavailability. In the simulated gastric juice, for the FDP variant, a decrease of the betalains content was observed, with a percentage of 35–40% after 30 min and around 40–45% after 120 min. In the simulated intestinal juice there was an increase in the release of the biologically active compounds with values between 3–3.7 times higher after 30 min of intestinal digestion and about 4–5 times higher after 120 min for all the studied products, thus suggesting a controlled release of these antioxidant compounds.

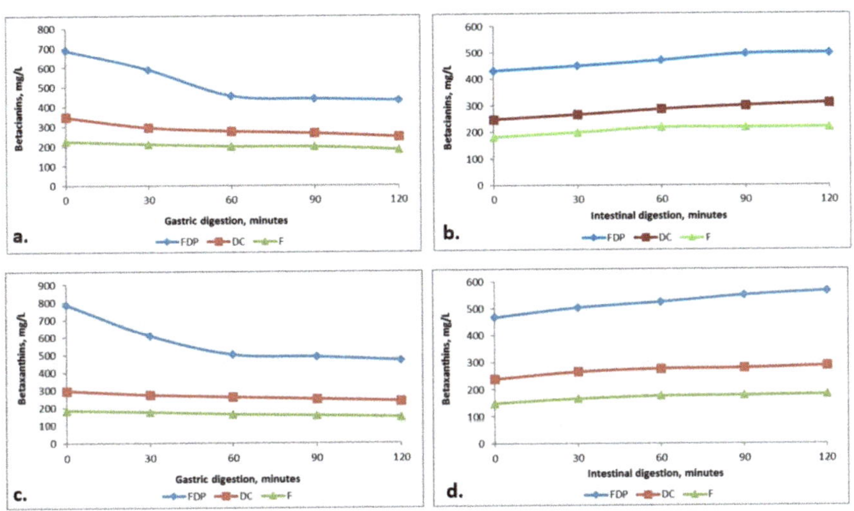

Figure 5. The digestibility of the beetroot products betacyanins (BC) and betaxanthins (BX) content during the gastric digestion (**a**,**c**) and the intestinal digestion (**b**,**d**).

4. Conclusions

The fresh, dried, or freeze-dried red beet samples, enriched with *Lb plantarum* BL3, showed a high antioxidant activity, an increased content of betalains and polyphenols, so their use as nutraceuticals is clearly justified. The obtained functional food products, based on beetroot enriched with lactic acid bacteria, were obtained for the daily consumption as dietary bioactive metabolites products that could provide a high content of bioactive molecules, mainly due to the fact that these antioxidant molecules showed great potential in scavenging free radicals, compounds that damage the healthy cells hence causing many diseases. The functional products designed as ready-to-eat single dosage (100 g of fresh or dried chips variants and 10 g of freeze-dried powder) can ensure a daily intake of 10^8–10^9 probiotic microorganisms, a concentration that is sufficient to achieve a probiotic action on the human organism. To the best of our knowledge, this is the first time when such food products were achieved, the study being of special interest both to the scientific community and to the manufacturers from the food or pharmaceutical industry.

Author Contributions: Conceptualization, V.B. and E.E.; investigation, V.B., M.C., C.A.B., D.G.A., A.C., E.E.; writing—original draft preparation, V.B., M.C., C.A.B., D.G.A., E.E.; writing—review and editing, V.B., G.E.B., E.E. All authors have read and agreed to the published version of the manuscript.

Funding: This research was funded by the project "Excellence, performance and competitiveness in the R&D&I activities at "Dunarea de Jos" University of Galati", acronym "EXPERT", financed by the Romanian Ministry of Research and Innovation, Programme 1—Development of the national research and development system, Sub-programme 1.2—Institutional Performance—Projects for financing excellence in R&D&I, Contract no. 14PFE/17.10.2018. The work of the Elena Enachi (E.E.) was supported by the project ANTREPRENORDOC, in the framework of Human Resources Development Operational Programme 2014–2020, financed from the European Social Fund under the contract number 36355/23.05.2019 HRD OP/380/6/13—SMIS Code: 123847.

Acknowledgments: The authors would like to thank for the technical support offered by the Integrated Center for Research, Expertise and Technological Transfer in Food Industry (BioAliment-TehnIA) and the MoRAS center, Grant POSCCE ID 1815, SMIS 48745 (www.moras.ugal.ro).

Conflicts of Interest: The authors declare no conflict of interest.

References

1. Šaponjac, V.T.; Čanadanović–Brunet, J.; Ćetković, G.; Jakišić, M.; Djilas, S.; Vulić, J.; Stajčić, S. Encapsulation of beetroot pomace extract: RSM optimization, storage and gastrointestinal stability. *Molecules* **2016**, *21*, 584. [CrossRef] [PubMed]

2. Georgiev, V.G.; Weber, J.; Kneschke, E.M.; Denev, P.N.; Bley, T.; Pavlov, A.I. Antioxidant activity and phenolic content of betalain extracts from intact plants and hairy root cultures of the red beetroot *Beta vulgaris* cv. Detroit dark red. *Plant Food Hum. Nutr.* **2010**, *65*, 105–111. [CrossRef] [PubMed]

3. Zielinska-Przyjemska, M.; Olejnik, A.; Dobrowolska-Zachwieja, A.; Grajek, W. In vitro effects of beetroot juice and chips on oxidative metabolism and apoptosis in neutrophils from obese individuals. *Phytother. Res.* **2009**, *23*, 49–55. [CrossRef] [PubMed]

4. Tesoriere, L.; Butera, D.; D'Arpa, D.; Di Gaudio, F.; Allegra, M.; Gentile, C.; Livrea, M.A. Increased resistance to oxidation of betalain-enriched human low-density lipoproteins. *Free Radic. Res. Commun.* **2003**, *37*, 689–696. [CrossRef]

5. Reddy, M.K.; Alexander-Lindo, R.L.; Nair, M.G. Relative inhibition of lipid peroxidation, cyclooxygenase enzymes, and human tumor cell proliferation by natural food colors. *J. Agric. Food Chem.* **2005**, *53*, 9268–9273. [CrossRef] [PubMed]

6. Wruss, J.; Waldenberger, G.; Huemer, S.; Uygun, P.; Lanzerstorfer, P.; Müller, U.; Höglinger, O.; Weghuber, J. Compositional characteristics of commercial beetroot products and beetroot juice prepared from seven beetroot varieties grown in Upper Austria. *J. Food Compos. Anal.* **2015**, *42*, 46–55. [CrossRef]

7. Gandia-Herrero, F.; Escribano, J.; Garcia-Carmona, F. Structural implications on color, fluorescence, and antiradical activity in betalains. *Planta* **2010**, *232*, 449–460. [CrossRef]

8. Stintzing, F.C.; Carle, R. Betalains in food: Occurrence, stability, and postharvest modifications. In *Food Colourants: Chemical and Functional Properties*; Socaciu, C., Ed.; CRC Press: Boca Raton, FL, USA, 2008; pp. 277–299.

9. Strack, D.; Vogt, T.; Schliemann, W. Recent advances in betalain research. *Phytochemistry* **2003**, *62*, 247–269. [CrossRef]

10. Tsai, P.J.; Sheu, C.H.; Wu, P.H.; Sun, Y.F. Thermal and pH Stability of betacyanin pigment of djulis (*Chenopodium formosanum*) in Taiwan and their relation to antioxidant activity. *J. Agric. Food Chem.* **2010**, *58*, 1020–1025. [CrossRef]

11. Gaertner, V.L.; Goldman, I.L. Pigment distribution and total dissolved solids of selected cycles of table beet from a recurrent selection program for increased pigment. *J. Am. Soc. Hortic. Sci.* **2005**, *130*, 424–433. [CrossRef]

12. Kazimierczak, R.; Hallmann, E.; Lipowski, J.; Drela, N.; Kowalik, A.; Pussa, T.; Matt, D.; Luik, A.; Gozdowski, D.; Rembiałkowska, E. Beetroot (*Beta vulgaris* L.) and naturally fermented beetroot juices from organic and conventional production: Metabolomics, antioxidant levels and anti-cancer activity. *J. Sci. Food Agric.* **2014**, *94*, 2618–2629. [CrossRef] [PubMed]

13. Jackson, M.S.; Bird, A.R.; Mc-Orist, A.I. Comparison of two selective media for the detection and enumeration of lactobacilli in human faeces. *J. Microbiol. Methods* **2002**, *51*, 313–321. [CrossRef]

14. Nguyen, T.D.T.; Kang, J.H.; Lee, M.S. Characterization of *Lactobacillus plantarum* PH04, a potential probiotic bacterium with cholesterol-lowering effects. *Int. J. Food Microbiol.* **2007**, *113*, 358–361. [CrossRef]

15. Asgharzadeh, F.; Tanomand, A.; Reza Ashoori, M.; Asgharzadeh, A.; Zarghami, N. Investigating the effects of *Lactobacillus casei* on some biochemical parameters in diabetic mice. *JEMDSA* **2017**, *22*, 47–50. [CrossRef]

16. Shori, A.B.; Baba, A.S. Antioxidant activity and inhibition of key enzymes linked to type-2 diabetes and hypertension by *Azadirachta indica*-yogurt. *J. Saudi Chem. Soc.* **2013**, *17*, 295–301. [CrossRef]

17. Yadav, H.; Jain, S.; Sinha, P.R. Antidiabetic effect of probiotic dahi containing *Lactobacillus acidophilus* and *Lactobacillus casei* in high fructose fed rats. *Nutrition* **2007**, *23*, 62–68. [CrossRef] [PubMed]

18. Yu, A.Q.; Li, L. The Potential Role of Probiotics in Cancer Prevention and Treatment. *Nutr. Cancer* **2016**, *68*, 535–544. [CrossRef] [PubMed]
19. Shah, N.P. Functional cultures and health benefits. *Int. Dairy J.* **2007**, *17*, 1262–1277. [CrossRef]
20. De LeBlanc, A.D.M.; Chaves, S.; Carmuega, E.; Weill, R.; Antóine, J.; Perdigón, G. Effect of long-term continuous consumption of fermented milk containing probiotic bacteria on mucosal immunity and the activity of peritoneal macrophages. *Immunobiology* **2008**, *213*, 97–108. [CrossRef] [PubMed]
21. Kassayová, M.; Bobrov, N.; Strojný, L.; Kisková, T.; Mikeš, J.; Demečková, V.; Bojková, B.; Péč, M.; Kubatka, P.; Bomba, A. Preventive Effects of Probiotic Bacteria *Lactobacillus plantarum* and Dietary Fiber in Chemically-induced Mammary Carcinogenesis. *Anticancer Res.* **2014**, *34*, 4969–4976. [PubMed]
22. Liu, C.F.; Pan, T.M. In vitro effects of lactic acid bacteria on cancer cell viability and antioxidant activity. *J. Food Drug Anal.* **2010**, *18*, 77–86.
23. Moreno, Y.; Collado, M.C.; Ferrús, M.A.; Cobo, J.M.; Hernández, E.; Hernández, M. Viability assessment of lactic acid bacteria in commercial dairy products stored at 4 °C using LIVE/DEAD® BacLightTM staining and conventional plate counts. *Int. J. Food Sci. Technol.* **2005**, *41*, 275–280. [CrossRef]
24. Neagu, C.; Barbu, V. Principal component analysis of the factors involved in the extraction of beetroot betalains. *J. Agroaliment. Process Technol.* **2014**, *20*, 311–318.
25. Bucur, L.; Țarălungă, G.; Schroder, V. The betalains content and antioxidant capacity of red beet (*Beta vulgaris* l. subsp. *vulgaris*) root. *Farmacia* **2016**, *64*, 198–201.
26. Stintzing, F.C.; Schieber, A.; Carle, R. Evaluation of colour properties and chemical quality parameters of cactus juices. *Eur. Food Res. Technol.* **2003**, *216*, 303–311. [CrossRef]
27. Chandran, J.; Nisha, P.; Singhal, R.S.; Pandit, A.B. Degradation of colour in beetroot (*Beta vulgaris* L.): A kinetics study. *J. Food Sci. Technol.* **2014**, *51*, 2678–2684. [CrossRef]
28. Yuan, C.; Du, L.; Jin, Z.; Xu, X. Storage stability and antioxidant activity of complex of astaxanthin with hydroxypropyl-β-cyclodextrin. *Carbohydr. Polym.* **2013**, *91*, 385–389. [CrossRef]
29. Turturică, M.; Stănciuc, N.; Bahrim, G.; Râpeanu, G. Effect of thermal treatment on phenolic compounds from plum (*Prunus domestica*) extracts—A kinetic study. *J. Food Eng.* **2016**, *171*, 200–207. [CrossRef]
30. Croitoru, C.; Mureşan, C.; Turturică, M.; Stănciuc, N.; Andronoiu, D.G.; Dumitraşcu, L.; Barbu, V.; Enachi, E. (Ioniţă); Horincar, G. (Parfene); Râpeanu, G. Improvement of Quality Properties and Shelf Life Stability of New Formulated Muffins Based on Black Rice. *Molecules* **2018**, *23*, 3047. [CrossRef]
31. Tripathi, M.K.; Giri, S.K. Probiotic functional foods: Survival of probiotics during processing and storage. *J. Funct. Foods* **2014**, *9*, 225–241. [CrossRef]
32. EU. A2 Date of publication of application: 14.04.93 Bulletin 93/15. Method of vegetable processing. Patent 0 536 851. Application number: 92203124.0; Applicant: Matforsk Norwegien Food Research Institute; Osloveien 1N-1329 As(NO).
33. Salas-Jara, M.J.; Ilabaca, A.; Vega, M.; García, A. Biofilm Forming *Lactobacillus*: New Challenges for the Development of Probiotics. *Microorganisms* **2016**, *4*, 35. [CrossRef] [PubMed]
34. Lewis, K. Riddle of biofilm resistance. *Antimicrob. Agents Chemother.* **2001**, *45*, 999–1007. [CrossRef] [PubMed]
35. Cui, L.; Niu, L.Y.; Li, D.-J.; Liu, C.-G.; Liu, Y.-P.; Liu, C.-J.; Song, J.-F. Effects of different drying methods on quality, bacterial viability and storage stability of probiotic enriched apple snacks. *J. Integr. Agric.* **2018**, *17*, 247–255. [CrossRef]
36. Kushwaha, R.; Kumar, V.; Vyas, G.; Kaur, J. Optimization of Different Variable for Eco-friendly Extraction of Betalains and Phytochemicals from Beetroot Pomace. *Waste Biomass Valorization* **2017**, *8*, 3–12. [CrossRef]
37. Molina, G.A.; Hernández-Martínez, A.R.; Cortez-Valadez, M.; García-Hernández, F.; Estevez, M. Effects of tetraethyl orthosilicate (TEOS) on the light and temperature stability of a pigment *from Beta vulgaris* and its potential food industry applications. *Molecules* **2014**, *19*, 17985–18002. [CrossRef]
38. Sekiguchi, H.; Ozeki, Y.; Sasaki, N. *Red Beet Biotechnology: Food and Pharmaceutical Applications*; Neelwarne, B., Ed.; Springer Science & Business Media: Boston, MA, USA, 2012; pp. 60–63.
39. Ravichandran, K.; Saw, N.M.M.T.; Mohdaly, A.A.; Gabr, A.M.; Kastell, A.; Riedel, H.; Cai, Z.; Knorr, D.; Smetanska, I. Impact of processing of red beet on betalain content and antioxidant activity. *Food Res. Int.* **2013**, *50*, 670–675. [CrossRef]
40. Slavov, A.; Karagyozov, V.; Denev, P.; Kratchanova, M.; Kratchanov, C. Antioxidant activity of red beet juices obtained after microwave and thermal pretreatments. *Czech J. Food Sci.* **2013**, *31*, 139–147. [CrossRef]

41. Roy, K.; Gullapalli, S.; Chaudhuri, U.R.; Chakraborty, R. The use of a natural colorant based on betalain in the manufacture of sweet products in India. *Int. J. Food Sci. Technol.* **2004**, *39*, 1087–1091. [CrossRef]

42. Gokhale, S.V.; Le, S.S. Dehydration of Red Beet Root (*Beta vulgaris*) by Hot Air Drying: Process Optimization and Mathematical Modeling. *Food Sci. Biotechnol.* **2011**, *20*, 955–964. [CrossRef]

43. Del Rio, D.; Rodriguez-Mateos, A.; Spencer, J.P.; Tognolini, M.; Borges, G.; Crozier, A. Dietary (poly)phenolics in human health: Structures, bioavailability, and evidence of protective effects against chronic diseases. *Antioxid. Redox Signal.* **2013**, *18*, 1818–1892. [CrossRef]

44. Espin, J.C.; González-Sarrías, A.; Tomás-Barberán, F.A. The gut microbiota: A key factor in the therapeutic effects of (poly)phenols. *Biochem. Pharmacol.* **2017**, *139*, 82–93. [CrossRef] [PubMed]

 *foods*

Article

Development of a Millet Starch Edible Film Containing Clove Essential Oil

Alaa G. Al-Hashimi [1], Altemimi B. Ammar [1,*] , Lakshmanan G. [2], Francesco Cacciola [3] and Naoufal Lakhssassi [4]

1. Food Science Department, College of Agriculture, University of Basrah, 61004 Basrah, Iraq; dr.alaagh@yahoo.co.uk
2. Central Research Laboratory, Meenakshi Academy of Higher Education and Research, 600078 Chennai, India; lakshmanang261988@gmail.com
3. Department of Biomedical and Dental Sciences and Morphofunctional Imaging, University of Messina, 98125 Messina, Italy; cacciolaf@unime.it
4. School of Agricultural Sciences, Southern Illinois University at Carbondale, Carbondale, IL 62901, USA; naoufal.lakhssassi@siu.edu
* Correspondence: ammaragr@siu.edu; Tel.: +96-477-356-40090

Received: 15 January 2020; Accepted: 11 February 2020; Published: 13 February 2020

Abstract: Medicinal plants contain various secondary metabolites. The present study analyzed the essential oil of buds from clove (Syzygium aromaticum L.; Family: Myrtaceae) using gas chromatography-mass spectrometry (GC-MS). GC-MS analysis showed the presence of six major phytoconstituents, such as eugenol (66.01%), caryophyllene (19.88%), caryophyllene oxide (5.80%), phenol, 2-methoxy-4-(2-propenyl)-acetate (4.55%), and humulene (3.75%). The effect of clove essential oils (CEO) at 0%, 1%, 2%, and 3% (w/w) on the mechanical and barrier properties of starch films was evaluated. The tensile strength (TS) and elongation (E) of films with clove essential oil were 6.25 ± 0.03 MPa and 5.67% ± 0.08%, respectively. The antioxidant activity of the films significantly increased the millet starch film and presented the lowest antioxidant activity (0.3%) at a 30 minute incubation for the control sample, while increasing CEO fraction in the starch film lead to an increase in antioxidant activity, and the 3% CEO combined film presented the highest antioxidant activity (15.96%) at 90 min incubation. This finding could be explained by the incorporation of clove oil containing antioxidant properties that significantly increased with the incorporation of CEO ($p < 0.05$). A zone of inhibition ranging from 16 to 27 mm in diameter was obtained when using a concentration of CEO ranging from 1% to 3%. We also observed the presence of an antimicrobial activity on several tested microorganism including *Escherichia coli*, *Pseudomonas aeruginosa*, *Enterobacter sp*, *Bacillus cereus*, *Staphylococcus aureus*, and *Trichoderma fungi*. Thus, the current study reveals the possibility of using a millet starch edible film as a preservation method.

Keywords: millet starch; edible film; Clove; GC-MS

1. Introduction

In recent years, several investigations focused on the development of eco-friendly, edible, and bio-degradable films using natural resources. Materials that are recognized as safe substances were used, such as lipids, proteins, and polysaccharides, in order to develop edible film and coatings [1]. These materials can be consumed and work effectively as a barrier layer between the food and the surrounding environment [2].

Worldwide, medicinal plants are used for folk medicine and also are consumed either directly or indirectly by humans for health benefits. The secondary metabolites are classified into four main groups, including the flavonoids, terpenoids, nitrogen-containing alkaloids, and sulfur-containing compounds. The plant derivatives have been reported to be effective in the treatment of multi-drug resistance cancer [3,4].

Edible films have several applications, including coloring agents, antimicrobial, and bioactive properties [5]. Moreover, encapsulating antimicrobial substances with edible films can improve the quality of food products and delay the growth of microorganisms [6]. Starch is mostly found in carbohydrates, which play an essential role in people's diets and exist in granular structures of different sizes and shapes in plants. Starches are composed of two glucan molecules, such as amylose and amylopectin. Starch is also known as biopolymer and it is used as edible film since it could be obtained from wide range of raw materials, in addition to the ability to form films and being produced cheaply. The main reason for developing films of starch is to prevent the changes of taste, color, flavor, and appearance in food products [7].

Approximately 33.692 million hectares of millet was cultivated worldwide, and its production reached 26,702.000 metric tons [8]. Other studies reported that about 33.5 million hectares was used to cultivate millet and about 35 million tons was produced across the world [9]. People with low income could use the millet as an alternative source of carbohydrates because it contains higher amounts of starch [10]. Starch is mainly present in pearl millet and represents 59% to 80% of the endosperm. Barnyard millet contains 66% starch, 15% protein, and 7% lipids, in addition to various micronutrients [11,12]. Starch containing high amylose content is considered as a raw material for edible films presenting good oxygen barrier properties [13]. Some studies reported that starch-based edible films possess the ability to transfer the water vapor, and for this reason, waxes, lipid additives, and essential oils were used in order to improve the hydrophobic fraction side of the film [14].

Essential oils play an important role in antimicrobial activities due to their valuable composition from phenols and terpenes [15]. The synergistic effects between the essential oils and their components can enhance the functional properties of edible films and thereby increase the shelf-life of food production, especially the food containing high fat levels. Earlier studies used essential oils as a potential source to preserve food from deterioration. For instance, using essential oils can cause many problems related to toxicity, intense aroma, and change in the sensory properties of food products [14].

Gas chromatography (GC) analysis is used to analyze biological material containing volatile constituents [16]. Mass spectrometry is a powerful analytical technique for the identification and quantification of known and unknown compounds in order to reveal the structure and chemical properties of molecules.

The biological activity of clove essential oil (CEO) was studied and was found to have the ability to work as an antioxidant, insecticidal, antifungal, and antibacterial agent [15]. In particular, it has been reported that CEO contains enormous amounts of bioactive compounds, such as triterpenoids, sesquiterpenes, and tannins. In addition, some studies reported that one of the main components, *viz.* eugenol (4-allyl-2-methoxyphenol), works effectively as antifungal activity agent [17,18]. Eugenol is used as a food additive and classified to be a safe substance according to the United States Food and Drug Administration (FDA) [19].

The aim of the present study was to study the effect of millet starch-based films and characterize the physical properties as well as antioxidant activity and antimicrobial of the millet starch (MS) films when incorporated with CEO.

2. Material and Methods

2.1. Materials

Millet, in addition to the source of starch and the clove (*Syzygium aromaticum*) variety used in this study, were purchased from the Basrah local market.

2.2. Starch Extraction

Millet starch was extracted according to the method of Bhupender et al. [20]. Millet grains were soaked in distilled water/sodium azide (0.01%) at ratio (1:2) for 24 h at 4 °C to reduce microbial growth. The excess water was drainage steeped and the washed grains were ground in a warming blender with sufficient water. The slurry was sieved on an 85-mesh nylon bolting cloth. The remaining parts (millet peels, germ, and endosperm) were again slurred with water to float off the germ and peels. The grinding, sieving, and regrinding processes of the remnant's endosperm particles were repeated until they were basically free from starch. The starch–protein slurry was centrifuged at 2000 rpm for 20 min. The supernatant was discarded, and the protein layer on the top of the starch was removed with a spatula. The starch was washed repeatedly by re-dispersing in distilled water and centrifuging until it appeared clean. The cleaned starch was air-dried on a glass plate for 12 h, re-dispersed in water, and wet-sieved through a 100-mesh screen. The starch passing through the screen was recovered by centrifugation (LMC, 3000) (2000 rpm, 20 min) and dried in a hot air oven at 40 °C.

2.3. Extraction of Essential Oil

The fresh buds of cloves were washed, and the wet samples were dried in a 30 °C ventilated drying oven. A total of 2 kg of cloves sample were mixed with 5 L distilled water for hydro-distillation by Clevenger apparatus (LG-6656-100, Wilmad, Ottaw, ON, Canada). The mixture was heated in a vertical hydro-distillation unit at 100 °C for 24 h. The CEO was separated from condensed vapor through an auto oil/water separator [21].

2.4. Gas Chromatography-Mass Spectroscopy (GC-MS) Analysis

A GC-MS analysis of essential clove oil was carried out using the GC-MS electron impact ionization (EI) method on gas chromatography (Shimadzu) coupled with a GC-MS QP2010 plus Mass Spectrometer (Tokyo, Japan) with an auto-sampler (AOC-20S) and an auto-injector (AOC-20i). A fused silica capillary column HP5-MS (30 m × 0.32 mm, film thickness 0.25μm) was used. One microliter of sample was injected into the capillary column. Helium was the carrier gas at constant pressure of 100Kpa. The flow rate was 1 ml/s and the injector's temperature was 250 °C. The column temperature starts at 50 °C, settles for 3 min, and then increases by 15 °C every minute until it reaches 250 °C and remains at this temperature for 5 min. The components of the CEO were identified by comparing the spectra with those of known compounds stored in the National Institute of Standards and Technology (NIST) library (2005).

2.5. Preparation of the Starch Edible Film

The edible films were prepared according to Resianingrum [22] with a few modifications. Briefly, 5 grams of millet starch were dissolved in 150 mL of distilled water. The starches were melted using a continuous heated stirrer at 75 °C until the solution gelatinized to allow leaching. A total of 2 mL of glycerol was used as plasticizer and then mixed and heated at 60 °C for 30 min. After the heating process was completed, the mixture was cooled down to 30 °C. Next, CEO at three different concentrations (1%, 2%, 3%) was added slowly to the solution with continuous stirring. Approximately 40 mL of the film starch solution was poured onto glass plates, and then fixed to remove the outer rim in all four outlines. The glass plates were covered with aluminum foil. These plates are left until the solution was tightened for 5 h of drying at 75 °C.

2.6. Film Thickness

The film thickness was measured using a micrometer to the nearest 0.01 mm of accuracy at five random positions of the films.

2.7. Mechanical Properties

Mechanical properties, tensile strength, and elongation were measured using a Universal Testing Machine COM-TEN testing machine 95T series model at the polymer research center/Basrah University. The tests were carried out according to the American Society for Testing and Materials (ASTM) [23].

2.8. Water Vapor Permeability

The permeability of films for water vapor was determined according to ASTM [24] with some modifications. The cylindrical cup slot was coated by the circular film samples and was well established using rubber bands. These sylider contained 50 g $CaCl_2$ (0% relative humidity, RH) to preserve an RH difference of 75% through the film. The cell was kept in a desiccator at 25 °C containing a saturated NaCl solution (75% RH). The weight of each penetration vessel was recorded after 24 h and the water vapor permeability (WVP) of the films was calculated using the following Equation (1):

$$WVP = \frac{\Delta W \times X}{t \times A \times \Delta P} \tag{1}$$

where WAP is the water vapor permeability (g.mm/m^2.day.kPa); ΔW is the weight gain by going down (g).

2.9. Oxygen Transmission Rate

The oxygen transmission rate (OTR) of the millet starch films incorporated with CEO was determined at 23 °C and 50% ± 1% RH according to ASTM [24]. The films were placed on an aluminum foil mask with an open area of 5 cm^2. The transported oxygen through the films was performed by the carrier gas (N_2/H_2) and the colometric sensor. Measurements were made on hourly affinity to reach the stable state of oxygen transport. The permeability coefficients in cc-μm/ (m2day atm) were calculated on the basis of the oxygen transmission rate in a steady state, taking into consideration the thickness water solubility of films.

Three discs of the films were cut into pieces with a 2 cm diameter, weighed, and submerged in 50 mL of distilled water; then, they were slowly moved and periodically agitated for 24 h at 25 °C. The dry mass content of the primary and final samples was determined by drying the samples at 105 °C for 24 h [25].

2.10. DPPH Radical Scavenging Activity

The radical scavenging activity of the millet starch films enriched with CEO was estimated according to the method described by Maizura et al. [26]. The antioxidant activity of the films was determined using the DPPH (2,2-diphenyl-1-picrylhydrazyl) free radical scavenging assay. Briefly, 3 mL of the film solution was blended with 1mL of 1 mM methanol solution of DPPH. The mixture was homogenized by a magnetic stirrer and incubated in the dark at an ambient temperature for 30 min. The absorbance was measured against Methanol (blank) at 517 nm, and the percentage of DPPH radical scavenging activity was achieved by the following Equation (2):

$$\text{DPPH scavenging effect \% } \frac{Abs\ DPPH - Abs\ Extracty}{Abs\ DPPH} \times 100 \tag{2}$$

where *Abs DPPH* is the blank absorbance value at 517 nm of the methanol solution of DPPH. Abs extract is the absorbance value at 517 nm for the films sample.

2.11. Antimicrobial Activity Test

Different isolated microorganisms were obtained from Basra University/the College of Agriculture/the Food Science Department and were used in this study, including *Esherichia coli*, *Staphylococcus aureus*, *Pseudomonas aeruginosa*, *Enterobacter sp.*, *Micrococcus roseus*, *B. cereus*, and the mold

Trichoderma to detect the antimicrobial activity of edible film incorporated with CEO. The films were tested for their inhibition against the target microorganisms according to Tooraj et al. [27]. The edible films were cut into 6-mm diameter discs and then put on nutrient agar plates, which were previously inoculated with 0.2 mL of inoculums containing approximately 10^5–10^6 CFU/ml of the bacteria. Then, the inoculum was spread in a circular motion until all the liquid was absorbed. The plates were then incubated at 37 °C for 24 h. As for *Trichoderma* sp, Malt extract agar petri dishes were prepared and seeded with the mold culture, then incubated at 22–25 °C for 2–5 days. Next, the plates were examined for a zone of inhibition on the film discs.

2.12. Statistical Analysis

The statistical analysis was carried out, employing the Statistical Package for the Social Sciences (SPSS) program using analysis of variance (ANOVA) to investigate the effect of CEO. The obtained result indicates a significant difference for the addition of the Cloves' essential oils and their concentrations.

3. Results and Discussion

3.1. Chemical Composition of Clove Oil

Using the steam distillation method with an average yield of 5.33%, GC-MS analysis (Figure 1) detected the presence of the chemical components and composition of CEO (*Syzygium aromaticum*), which are shown in Table 1. The GC-MS analysis showed the presence of various secondary metabolite, namely eugenol (66.01%), caryophyllene (19.88%), caryophyllene oxide (5.80%), phenol, 2-methoxy-4-(2-propenyl)-acetate (4.55), and humulene (3.75).

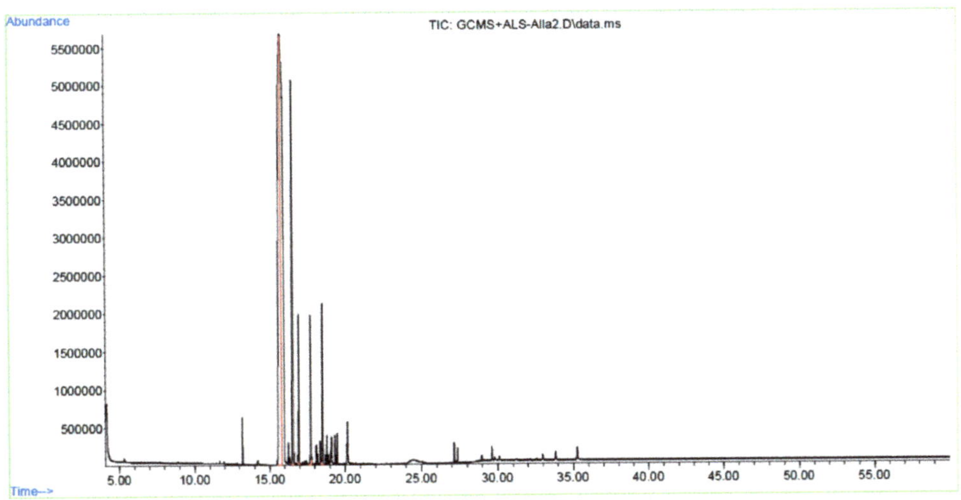

Figure 1. A typical gas chromatogram of the constituents of CEO.

Table 1. Percentage composition of volatile constituents of clove essential oils (CEO).

No.	Compound	t_R (min)	Molecular Weight	Composition %
1	Eugenol	15.80	164.2	66.01 ± 0.23
2	Caryophyllene	16.59	204.36	19.88 ± 0.11
3	Humulene	16.96	204.35	3.75 ± 0.65
4	Phenol, 2-methoxy-4-(2-propenyl)-acetate	17.74	206.23	4.55 ± 0.17
5	Caryophyllene oxide	18.53	220.35	5.80 ± 0.21

The obtained results are in agreement with Chaieb et al. [28] who reported eugenol (88.58%) as the major compound, followed by eugenol acetate (5.62%), and β caryophyllene (1.39%). However, 2-heptanone, ethyl hexanoate, humulenol, α-humulene, calacorene, and calamenene were detected but with very little amount (<1%). Similar results were also reported by Uddin et al. [29], where 3-Allyl-6-methoxyphenol *viz. m*-eugenol (69.44%) was detected as the major constituent, followed by eugenol acetate (10.79%), 4-hydroxy-4-mehtylpentan-2-one viz. Tyranton (7.78%), caryophyllene (6.80%), and 1,4,7-cycloundecatriene, 1,5,9,9-tetramethyl-,Z,Z,Z.

The variation of physical and chemical properties of CEO depends on several factors, including the tissue and origin of the plant, growing season, weather, harvest time, and air humidity. Another important factor is the time between the raw material harvest and oil production. More than 100 ingredients of CEO have been detected worldwide [30].

Eugenol ($C_{10}H_{12}O_2$) (Figure S1) is a volatile phenolic component and is the primary component found in the extracted clove buds' essential oil [31]. It has a molecular weight of 164.2 g/mol with a peculiar spicy aroma [32]. CEO also presents other components of terpenes, e.g., the ester form of eugenol (Phenol, 2-methoxy-4-(2-propenyl)-acetate) (Figure S2), with documented beneficial properties, including antibacterial [33], antifungal [34], antioxidant [35], and anti-inflammatory [36].

(−)-β-caryophyllene (Figure S3), is a common ubiquitous bicyclic sesquiterpene in nature, and is composed of many essential oils including the extracted oil from the stalk and buds of *Syzygium aromaticum* (cloves) [37].

(−)-β-caryophyllene and its derivative oxide have a severe woody aroma and are utilized as cosmetics in food manufacturing. These two natural components are used as flavors by the Food and Drug Administration (FDA) and by the European Food Safety Authority (EFSA) because of their low toxicity and low water solubility [38]. These compounds were also known to have antimicrobial and antioxidant properties [39], an anti-inflammatory activity against carrageenan and prostaglandin [40], and enhance skin penetration [41].

Humulene (Figure S4), which is known by other designations as α-humulene or α-caryophyllene, is a ring-opened isomer of β-caryophyllene oxide (Figure S5). It occurs in nature as a monocyclic sesquiterpene ($C_{15}H_{24}$), including an 11-membered ring, and it is composed of three isoprene units, including three not associated C=C double bonds, two of which are being replaced by a triple bonds [42]. The sesquiterpene hydrocarbon plays a very important role as an antimicrobial activity [43]. It also has a strong anti-inflammatory activity equal to dexamethasone [44].

3.2. Physical, Mechanical, and Barrier Properties of Millet Starch Edible Films Incorporated with Cloves' Essential Oil.

The importance of the mechanical properties of the edible film is due to its beneficial effects during trading and storage; therefore, it is one of the standards that specify the firmness of the film and its ability to promote food safety in food packaging applications, as films need to be strong, durable, and flexible [45].

In the application of polymer film packaging, thickness is an important aspect that requires special attention regarding the material design. The biomass thickness significantly affects other important properties, such as strength, elasticity, and moisture content. The addition of CEOs on a millet starch-based edible film shows a significant effect on film thickness. Table 2 indicates the highest oil concentrations that created an increase in thickness. Thus, the edible film containing 3% CEO has the highest rate of thickness 0.150 mm. This increase is due to the ultimate ingredient component of the edible film composition, as the film will be very thick and has more thickness than the other formulation [46]. According to Nugroho et al. [47], the increase of material concentration using different components for edible files will result in an increase in film thickness, which is coherent with the results obtained in the current study.

Table 2. Tensile strength (TS), thickness (TH), elongation at break (E), water vapor permeability (WVP), oxygen permeability coefficient (PO_2), and the solubility (S) of millet starch films incorporated with different contents of CEO.

Film	TH (mm)	TS(MPa)	E (%)	WVP	PO_2	S (%)
Control	0.120 ± 0.003	10.52 ± 0.05	9.3 ± 0.05	6.92 ± 0.088	19.70 ± 0.57	30.40 ± 0.3
MS- films (1% CEO)	0.130 ± 0.008	8.60 ± 0.08	7.43 ± 0.01	9.67 ± 0.088	24.50 ± 0.1	28.67 ± 0.14
MS- films (2% CEO)	0.135 ± 0.001	7.16 ± 0.05	6.25 ± 0.05	11.33 ± 0.033	26.25 ± 0.57	27.50 ± 0.8
MS- films (3% CEO)	0.150 ± 0.008	6.25 ± 0.03	5.67 ± 0.08	12.52 ± 0.08	28.87 ± 0.8	27.13 ± 0.145

As shown in Table 2, the content of the CEO has an effect on the tensile strength of the edible film. The strength length of the standard edible (without CEO) film was significantly different from the films that were enriched with the CEO. The TS of starch edible films incorporated with CEO decreased gradually as the concentration of CEO increased, ranging from 8.60 to 4.40 MPa. The decreasing value of TS was due to the interactions between the starch molecules [48]. It has been reported that essential oils plasticizing capability leads to the reduction in TS [49]. This result was in agreement with the result reported by Maizura et al. [26] who indicated a reduction in the tensile strength of films made from the starch–alginate film, in which different concentrations of lemon oil were added. Further, compared to control films, the lower mechanical resistance of films containing CEO can be explained by the composition of emulsion films. In those structures, the lipid molecules filled the protein matrix and the interactions between lipid and polar molecules occurred, which seemed to be weaker than the polar molecules of the control films.

Elongation is defined as the percentage change in the length of the membrane from its original length. The results show a change in the elongation of edible films when changing the concentration of clove oil, where the elongation was, in the case of the native starch edible film, 9.3%. However, it became 5.67% using CEO within a concentration of 3%. The potential cause of this increase is linked to the fact that the water-resistant surfactant has affected the hydrogen interstitially of the molecules within starch or starch in water [50]. Increasing the elongation at break is considered a positive affect regarding the flexibility of films, especially for those materials used as a package. This phenomenon is due to the presence of oils that play a role as a plasticizer or a lubricant in the hydrocolloid matrix. The obtained results reveal that the starch millet films enriched with CEO were less resistant with less extension than the film without CEO. This could be explained by the fact that lipids are incapable of maintaining a cohesive and uninterrupted matrix [51].

Low tensile and elongation values were the most common results of essential oils incorporation in bio-polymer matrices. The results of tensile strength and elongation were in agreement with Souza et al. [52], who found that the (*TS*) and (*E*) of films with cinnamon essential oil were 1.05 ± 0.16 MPa and 191.27% ± 22.62%, respectively. A loss of macromolecular mobility was obtained due to the increase in the cinnamon essential oil, glycerol, and emulsifier contents that impacted the TS and E of the films. The obtained data show that the control films (without essential oil) presented higher TS (3.96 ± 0.60) MPa but lower E (123.61% ± 19.57%) [53].

Water vapor permeability is one of the most important factors in the quality of food packaging materials. The applicable film must be able to avoid, or at least reduce water transfer between environment and food. As shown in Table 2, the WVP of the edible films increased proportionally with the concentration of CEO from 1% to 3% (9.67, 12.52 g mm^{-2} d^{-1} KPa^{-1}) when compared to the control film, which was 6.92× (g mm^{-2} d^{-1} KPa^{-1}). The water vapor permeability of edible films was reduced when the hydrophobic component of the edible film was increased. The hydrophobic compounds can play an important role in reducing the surface tension of the solution. The permeability of membranes made of starch regarding water vapor is influenced by several factors, including temperature, the thickness of the membrane, and the content of the additives [54].

The obtained low WVP rate in starch films on CEO may be due to hydrogen and covalent interactions between the starch network and the polyphenolics compounds. These reactions probably minimize the accessibility of hydrogen groups to form a hydrophilic link with water, resulting in a reduction in the affinity of the film [55].

It is desirable to have a limited exposure of food to oxygen because it can lead to oxidation in addition to sensory changes (odor, color, flavor, and texture) and the loss of nutrition [56]. The results of oxygen permeability of the millet starch edible film with and without CEO are shown in the Table 2. The increased content of CEO caused an increase in the oxygen permeability coefficient (PO_2) from (24.50 to 28.78) $g\ s^{-1}\ m^{-1}\ Pa^{-1} \times 10^{-10}$.

The composition of the matrix affects the proliferation of gas molecules through the polymer, resulting in significant variations in gas transmission. The primary CEO modifies the performance of the barrier, which is associated with the compatibility of starch and oil, eventually leading to the permeation of the effective gas molecule through films [57]. The effect of adding fat on the microstructure of the edible films is a critical factor in barrier efficiency. The physical state of essential oils and their distribution in the polymer matrix impacted significantly the microstructure of films. The liquid state of essential oils can improve molecular movement and promotes the transport of gas molecules [58]. In fact, the authors investigated properties of starch-based edible films incorporated with oregano and black cumin essential oil and indicate that water vapor barrier properties decreased proportionally with essential oils addition [58].

The solubility in water of the prepared starch films with/without CEO is presented in Table 2. The solubility value of the control film was 30.40%, while solubility of the films with CEO addition decreased from 28.67% to 27.13% with 1% and 3% CEO, respectively. When EO was added in starch film, solubility in water decreased. The engagement of EO in the structure of starch and interaction between the CEO and the hydroxyl groups of starch can increase the hydrophobic nature of the films. Therefore, the availability of hydroxyl groups and its interaction with water molecules was reduced and led to less solubility. Perhaps the decrease of starch film with CEO solubility is due to the formation of amylose-lipid complexes with a tight helical structure due to the formed links with the hydrophobic hole [59].

3.3. Free Radical Scavenging Activity

A DPPH scavenging assay was employed to elucidate the antioxidant activity of the millet starch films with and without CEO (Table 3). As the concentration of essential oil increased, the DPPH scavenging activity of the films was significantly increased in the millet starch film and presented the lowest antioxidant activity (0.3%) at 30 minute incubation for the control sample, while increasing the CEO fraction in the starch film lead to an increase of antioxidant activity and 3% CEO combined film presenting the highest antioxidant activity (15.96%) at 90 min incubation, which may be due to the presence of eugenol shown as the principal component of buds oil from clove. This was in agreement with the reported highest antioxidant activity of films containing additional eugenol microcapsules [60]. In fact, it has been shown that the addition of eugenol microcapsules containing oleic acid promoted the eugenol retention in the starch matrix during film formation and impacted positively the antioxidant activity. When films were developed with encapsulated eugenol powder containing lecithin and oleic acid, low and constant values of peroxide index, conjugated dienes, and trienes were observed, resulting in a high and effective prevention of sunflower oil oxidation even over seven weeks of storage [60].

Table 3. Antioxidant activity of millet starch (MS) edible film incorporated with CEO.

Film	% Inhibition of DPPH ± SD	
	30 Minutes Incubation	90 Minutes Incubation
Control	0.3 ± 0.100	0.7 ± 0.057
MS-films (1% CEO)	13.88 ± 0.075	36.57 ± 0.337
MS-films (2% CEO)	17.50 ± 0.100	57.34 ± 0.020
MS-films (3% CEO)	23.22 ± 0.890	85.96 ± 0.14

The antioxidant activities of these biodegradable films are related to the type and concentration of essential oils. This significant antioxidant capacity of CEO could be attributed to a higher content of phenolic components, such as Eugenol, Caryophyllene, Humulene, and Caryophyllene oxide. Shojaee-Aliabadi et al. [61] reported that phenol compounds are accountable for antioxidant activity in EO. Fernandes de Oliveira et al. [62] also indicated that phenolic compounds are receptors for free radicals by breaking the chain oxidation reactions or by metal clawing, which can be an indicator of the antioxidant capacity of the CEO.

A significant amount of antioxidants in cloves belong to the high content of phenolic compounds, such as eugenol and eugenyl acetate, and their ability to donate hydrogen (which is an effective radical scavenger) [63], ISO-eugenol, and caryophyllene [64], but with lower amounts of benzyl alcohol, chavicol, acetyl salicylate, and humulenes [65].

This finding was in agreement with Dashipour et al. [66], who reported that the antioxidant activity of the carboxymethylcellulose (CMC) film without EO was 0.32%, while the antioxidant activity of CMC film with CEO was 71.76%.

The antibacterial activity of *Syzygium aromaticum* showed maximum activity at different concentrations. The starch edible film containing different concentrations of oils of *Syzygium aromaticum* were screened for antimicrobial activity against five standard bacteria species: one Gram-positive bacteria *Staphylococcus aureus*, and four Gram-negative bacterial strains, including *Escherichia coli*, *Pseudomonas aeruginosa, Enterobacter sp*, and *B. cereus*, as well as one standard fungal strain *Trichoderma*.

The antibacterial activity of the AgNPs was dependent on the concentration. Elevated levels of AgNPs exhibited more inhibitory activity, impacting the growth of microbes. Different results of inhibition were obtained from the millet starch-based film incorporated with different concentrations of cloves' essential oils and for each microorganism studied (Table 4 and Figure 2). The statistical analysis indicated significant differences among the antimicrobial activity of films with different cloves' essential oil contents ($p < 0.05$). Certainly, there was no affect against microorganisms for the edible film without CEO (control). The films exhibited strong antimicrobial effects against all tested bacterial strains, including *Escherichia coli, Pseudomonas aeruginosa, Enterobacter sp, B. cereus, Staphylococcus aureus*, and *Trichoderma fungi*. Additionally, 3% of clove oils showed a zone of inhibition ranging from 16–27 mm in diameter. These results show that starch-based edible films incorporated with different CEO amounts have inhibitory efficiency against both positive and negative bacteria, while the inhibitory effect increases proportionally with CEO amount. However, the inhibitory activity of clove oil due to the presence of several constituents was mainly observed in eugenol and eugenyl acetate and β- caryophyllene. These components can change protein structure and the phospholipids of cell membranes by affecting their permeability [67]. The hydrophobic criteria of essential oils interact with the lipid structure, such as Gram-negative bacteria cell membrane, mitochondria, and most intracellular component, which lead to damaging the cell structure, leaching, ion exchange, breathing inhibition, and finally causing cell death [68,69]. These reported phenotypes were coherent with our findings. The positive strain *Staphylococcus aureus* has the lowest sensitivity toward the films' cloves oil components, especially eugenol, which is the main factor inhibiting fungal activity due of the lyses of spores and micelles [70,71].

Table 4. Antimicrobial activity of the MS edible film incorporated with CEO.

No.	Standard Microorganisms	Zone of Inhibition (mm)			
		Concentration (%)			
		Control (0)	1	2	3
	Tasted Bacteria				
1	*Escherichia coli*	0	16 ± 0.13	18 ± 0.16	23 ± 0.43
2	*Pseudomonas aeruginosa*	0	12 ± 0.15	14 ± 0.13	24 ± 0.32
3	*Enterobacter sp.*	0	14 ± 0.16	16 ± 0.11	27 ± 0.81
4	*B.cereus*	0	11 ± 0.34	12 ± 0.84	20 ± 0.52
5	*Staphylococcus aureus*	0	10 ± 0.58	11 ± 0.52	18 ± 0.95
	Tasted Fungi				
	Trichoderma	**0**	13 ± 0.65	27 ± 0.32	14 ± 0.76

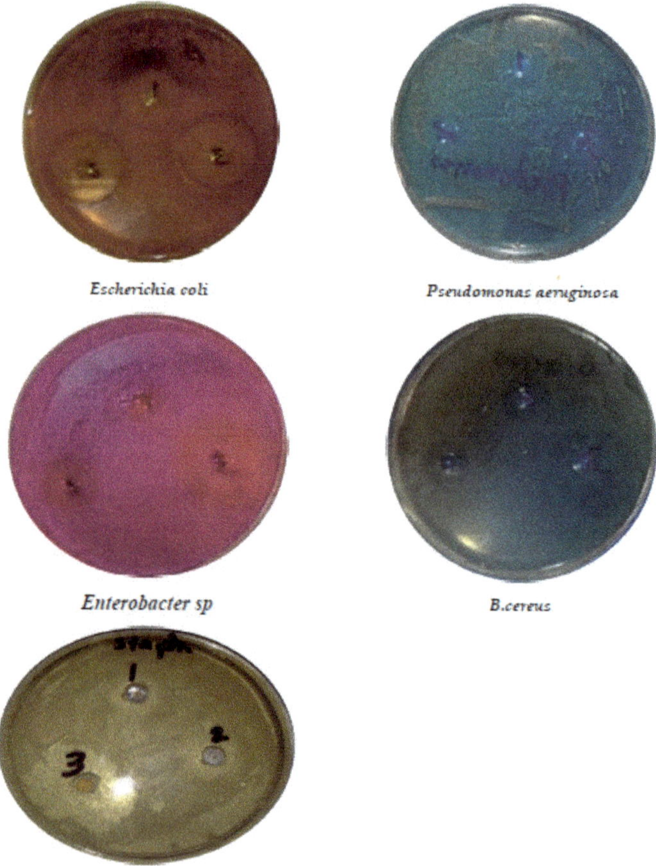

Figure 2. Antimicrobial activity of the MS edible film incorporated with CEO.

4. Conclusions

Edible films play an important role in the revelation of packing and keeping the environment safe. Millet starch edible films with CEOs containing polyphenolic components provide new ways to enhance antioxidant and microbial safety and extend the shelf life of foods. The present study showed the presence of significant differences in the mechanical and barrier characteristic since the concentration of clove oil varied from 1–3%. The active volatile and semi-volatile compounds were ascertained by GC-MS analysis. The results also showed that the presence of significant amounts of antioxidants in cloves was due to the high content of phenolic compounds. The current study also revealed that the engagement of EO in the structure of starch and the interaction between the CEO and the hydroxyl groups of starch can increase the hydrophobic nature of the films. The decrease in solubility is most probably due to the availability of hydroxyl groups, which reduces its interaction with water molecules. The study also demonstrated that the major phytoconstituent of the essential oil of clove buds was the eugenol (66.01%). The obtained results are coherent with other studies that showed that encapsulated eugenol modified the film microstructure and yielded less stretchable films with a reduced water affinity, transparency, and oxygen permeability when compared to films formulated with non-encapsulated eugenol [60].

Supplementary Materials: The following are available online at http://www.mdpi.com/2304-8158/9/2/184/s1, Figure S1: MS spectrum of Eugenol; Figure S2: MS spectrum of Eugenol acetate; Figure S3: MS spectrum of caryophyllene; Figure S4: MS spectrum of humulene; Figure S5: MS spectrum of caryophyllene oxide.

Author Contributions: Investigation, F.C.; Methodology, A.G.A.-H.; Supervision, A.B.A.; Writing—original draft, L.G.; Writing—review & editing, N.L. All authors have read and agreed to the published version of the manuscript.

Funding: This research received no external funding

Acknowledgments: The authors would like to thank Food science department, Agriculture College, University of Basrah for providing facilities and equipment.

Conflicts of Interest: There is no conflict of interest.

References

1. Cho, S.Y.; Rhee, C. Mechanical properties and water vapor permeability of edible films made from fractionated soy proteins with ultrafiltration. *LWT Food Sci. Technol.* **2004**, *37*, 833–839. [CrossRef]

2. Garcia, M.A.; Martino, M.N.; Zaritzky, N.E. Lipid addition to improve barrier properties of edible starch-based films and coatings. *J. Food Sci.* **2000**, *65*, 941–944. [CrossRef]

3. Nirmala, M.J.; Samundeeswari, A.; Sankar, P.D. Natural plant resources in anti-cancer therapy—A review. *Res. Plant Biol.* **2011**, *1*, 1–14.

4. Bonilla, J.; Atarés, L.; Vargas, M.; Chiralt, A. Effect of essential oils and homogenization conditions on properties of chitosan-based films. *Food Hydrocoll.* **2012**, *26*, 9–16. [CrossRef]

5. Cagri, A.; Ustunol, Z.; Ryser, E.T. Antimicrobial edible films and coatings. *J. Food Protect.* **2004**, *67*, 833–848. [CrossRef] [PubMed]

6. Ko, S.; Janes, M.; Hettiarachchy, N.; Johnson, M.G. Physical and chemical properties of edible films containing nisin and their action against Listeria monocytogenes. *J. Food Sci.* **2001**, *66*, 1006–1011. [CrossRef]

7. Chiumarelli, M.; Hubinger, M.D. Stability, solubility, mechanical and barrier properties of cassava starch–Carnauba wax edible coatings to preserve fresh-cut apples. *Food Hydrocoll.* **2012**, *28*, 59–67. [CrossRef]

8. Food and Agricultural Organisation. *Countrystat, Economic*; Food and Agricultural Organisation: Rome, Italy, 2009.

9. Taylor, J.R. Sorghum and Millets: Taxonomy, History, Distribution, and Production. In *Sorghum and Millets*, 2nd ed.; AACC International Press: Washington, DC, USA, 2019; pp. 1–21.

10. Freeman, J.E.; Bocan, B.J. Pearl millet: A potential crop for wet milling. 1973. *Cereal Sci. Today* **1973**, *18*, 69–73.

11. Kim, J.Y.; Jang, K.C.; Park, B.R.; Han, S.I.; Choi, K.J.; Kim, S.Y.; Hwang, J. Physicochemical and antioxidative properties of selected barnyard millet (Echinochloa utilis) species in Korea. *Food Sci. Biotechnol.* **2011**, *20*, 461–469. [CrossRef]

12. Krishnakumari, S.; Thayumanavan, B. Content of starch and sugars and in vitro digestion of starch by α-amylase in five minor millets. *Plant Food. Hum. Nutr.* **1995**, *48*, 327–333. [CrossRef]

13. Weiwei, L.; Juan, X.; Beijiu, C.; Suwen, Z.; Qing, M.; Huan, M. Anaerobic biodegradation, physical and structural properties of normal and high-amylose maize starch films. *Int. J. Agric. Biol. Eng.* **2016**, *9*, 184–193.

14. Sánchez-González, L.; Vargas, M.; González-Martínez, C.; Chiralt, A.; Cháfer, M. Use of essential oils in bioactive edible coatings: A review. *Food Eng. Rev.* **2011**, *3*, 1–16. [CrossRef]

15. Burt, S. Essential oils: Their antibacterial properties and potential applications in foods-a review. *Int. J. Food Microbiol.* **2004**, *94*, 223–253. [CrossRef] [PubMed]

16. Alsaad, A.J.A.; Altemimi, A.B.; Aziz, S.N.; Lakhssassi, N. Extraction and Identification of Cactus Opuntia dillenii Seed Oil and its Added Value for Human Health Benefits. *Pharmacogn. J.* **2019**, *11*, 1–9. [CrossRef]

17. Lee, K.G.; Shibamoto, T. Antioxidant property of aroma extract isolated from clove buds. *Syzygium aromaticum* *(L.) Merr. et Perry. Food Chem.* **2001**, *74*, 443–448.

18. Miyazawa, M.; Hisama, M. Antimutagenic activity of phenylpropanoids from clove (*Syzygium aromaticum*). *J. Agric. Food Chem.* **2003**, *51*, 6413–6422. [CrossRef]

19. Gülçin, İ.; Elmastaş, M.; Aboul-Enein, H.Y. Antioxidant activity of clove oil-A powerful antioxidant source. *Arab. J. Chem.* **2012**, *5*, 489–499. [CrossRef]

20. Bhupender, S.K.; Rajneesh, B.; Baljeet, S.Y. Physicochemical, functional, thermal and pasting properties of starches isolated from pearl millet cultivars. *Int. Food Res. J.* **2013**, *20*, 1555–1561.

21. Natta, L.; Orapin, K.; Krittika, N.; Pantip, B. Essential oil from five Zingiberaceae for anti food-borne bacteria. *Int. Food Res. J.* **2008**, *15*, 337–346.

22. Resianingrum, R.; Atmaka, W.; Khasanah, L.U.; Kawuiji, K.; Utami, R.; Praseptiangga, D. Characterization of cassava starch-based edible film enriched with lemongrass oil (*Cymbopogon citratus*). *Nusantara Biosci.* **2016**, *8*, 278–282. [CrossRef]

23. ASTM. Standard Test Method for Oxygen Gas Transmission Rate through Plastic Film and Sheeting Using Coulometric Sensor. In *Annual Book of ASTM Standards*; ASTM Publisher: Chicago, IL, USA; Philadelphia, PA, USA, 1986; pp. 387–393.

24. ASTM. Standard Test Method for Water Vapor Transmission of Materials. Designation E96-95. In *Annual Book of ASTM Standards*; American Society for Testing and Materials: Philadelphia, PA, USA, 1995.

25. Colla, E.; do Amaral Sobral, P.J.; Menegalli, F.C. Amaranthus cruentus flour edible films: Influence of stearic acid addition, plasticizer concentration, and emulsion stirring speed on water vapor permeability and mechanical properties. *J. Agric. Food Chem.* **2006**, *54*, 6645–6653. [CrossRef] [PubMed]

26. Maizura, M.; Fazilah, A.; Norziah, M.H.; Karim, A.A. Antibacterial activity and mechanical properties of partially hydrolyzed sago starch–alginate edible film containing lemongrass oil. *J. Food Sci.* **2007**, *72*, C324–C330. [CrossRef] [PubMed]

27. Tooraj, M.; Hossein, T.; Sayed, M.; Abdol, R. Antibacterial, antioxidant and optical properties of edible starch-chitosan composite film containing Thymus kotschyanus essential oil. *Vet. Res. Forum.* **2012**, *3*, 167–173.

28. Chaieb, K.; Hajlaoui, H.; Zmantar, T.; Kahla-Nakbi, A.B.; Rouabhia, M.; Mahdouani, K.; Bakhrouf, A. The chemical composition and biological activity of clove essential oil, Eugenia caryophyllata (*Syzigium aromaticum* L. Myrtaceae): A short review. *Phytother. Res.* **2007**, *21*, 501–506. [CrossRef]

29. Uddin, M.A.; Shahinuzzaman, M.; Rana, M.S.; Yaakob, Z. Study of chemical composition and medicinal properties of volatile oil from clove buds (*Eugenia caryophyllus*). *Int. J. Pharm. Sci. Res.* **2017**, *8*, 895–899.

30. Nowak, K.; Ogonowski, J.; Jaworska, M.; Grzesik, K. Clove oil-properties and applications. *Chemik* **2012**, *66*, 145–152.

31. Barceloux, D.G. *Medical Toxicology of Natural Substances: Foods, Fungi, Medicinal Herbs, Plants, and Venomous Animals*; John Wiley & Sons: Hoboken, NJ, USA, 2008; pp. 619–621.

32. Oyedemi, S.O.; Okoh, A.I.; Mabinya, L.V.; Pirochenva, G.; Afolayan, A.J. The proposed mechanism of bactericidal action of eugenol, α-terpineol and γ-terpinene against Listeria monocytogenes, Streptococcus pyogenes, Proteus vulgaris and Escherichia coli. *Afr. J. Biotechnol.* **2009**, *8*, 1280–1286.

33. Moon, S.; Kim, H.; Cha, J. Synergistic effect between clove oil and its major compounds and antibiotics against oral bacteria. *Arch. Oral Biol.* **2011**, *56*, 907–916. [CrossRef]

34. Rana, I.; Rana, A.; Rajak, R. Evaluation of antifungal activity in essential oil of the *Syzygium aromaticum* (L.) by extraction, purification and analysis of its main component eugenol. *Braz. J. Microbiol.* **2011**, *42*, 1269–1277. [CrossRef]

35. Vanin, A.; Orlando, T.; Piazza, P.; Puton, M.; Cansian, L.; Oliveora, D.; Paroul, N. Antimicrobial and antioxidant activities of clove essential oil and eugenyl acetate produced by enzymatic esterification. *Appl. Biochem. Biotechnol.* **2014**, *174*, 1286–1298. [CrossRef]

36. Daniel, A.N.; Sartoretto, S.M.; Schmidt, G.; Caparroz-Assef, S.M.; Bersani-Amado, C.A.; Cuman, R.K.N. Anti-inflammatory and antinociceptive activities A of eugenol essential oil in experimental animal models. *Rev. Brasil. Farmacogn.* **2009**, *19*, 212–217. [CrossRef]

37. Ghelardini, C.; Galeotti, N.; Mannelli, L.D.C.; Mazzanti, G.; Bartolini, A. Local anaesthetic activity of β-caryophyllene. *Il Farmaco* **2011**, *56*, 387–389. [CrossRef]

38. Sarpietro, M.G.; Di Sotto, A.; Accolla, M.L.; Castelli, F. Interaction of β-caryophyllene and β-caryophyllene oxide with phospholipid bilayers: Differential scanning calorimetry study. *Thermochim. Acta* **2015**, *600*, 28–34. [CrossRef]

39. Dahham, S.; Tabana, Y.; Iqbal, M.; Ahamed, M.; Ezzat, M.; Majid, A.; Majid, A. The anticancer, antioxidant and antimicrobial properties of the sesquiterpene β-caryophyllene from the essential oil of Aquilaria crassna. *Molecules* **2015**, *20*, 11808–11829. [CrossRef] [PubMed]

40. Agarwal, R.B.; Rangari, V.D. Phytochemical investigation and evaluation of anti-inflammatory and anti-arthritic activities of essential oil of Strobilanthus ixiocephala Benth. *Indian J Exp. Biol.* **2003**, *41*, 890–894. [PubMed]

41. Lucca, L.G.; de Matos, S.P.; Borille, B.T.; Dias, D.D.O.; Teixeira, H.F.; Veiga, V.F., Jr.; Koester, L.S. Determination of β-caryophyllene skin permeation/retention from crude copaiba oil (Copaifera multijuga Hayne) and respective oil-based nanoemulsion using a novel HS-GC/MS method. *J. Pharm. Biomed. Anal.* **2015**, *104*, 144–148. [CrossRef]

42. Tinseth, G. *Hop Aroma and Flavor, January/February, Brewing Techniques*; Brewers Publications: Boulder, CO, USA, 1993.

43. Rahman, M.M.; Garvey, M.; Piddock, L.J.; Gibbons, S. Antibacterial terpenes from the oleo-resin of Commiphora molmol (Engl.). *Phytother. Res.* **2008**, *22*, 1356–1360. [CrossRef]

44. Fernandes, E.S.; Passos, G.F.; Medeiros, R.; da Cunha, F.M.; Ferreira, J.; Campos, M.M.; Calixto, J.B. Anti-inflammatory effects of compounds alpha-humulene and (−)-trans-caryophyllene isolated from the essential oil of Cordia verbenacea. *Eur. J. Pharmacol.* **2007**, *569*, 228–236. [CrossRef]

45. Ghanbarzadeh, B.; Almasi, H. Physical properties of edible emulsified films based on carboxymethyl cellulose and oleic acid. *Int. J. Biol. Macromol.* **2011**, *48*, 44–49. [CrossRef]

46. Prasetyaningrum, A.; Rokhati, N.; Kinasih, D.N.; Wardhani, F.D.N. Karakterisasi bioactive edible film dari komposit alginat dan lilin lebah sebagai bahan pengemas makanan biodegradable. *Seminar Rekayasa Kimia Dan Proses* **2010**, *2*, 1411–4216.

47. Nugroho, A.A.; Basito, B.; Anandito, R.B.K. Kajian Pembuatan Edible Film Tapioka dengan Pengaruh Penambahan Pektin Beberapa Jenis Kulit Pisang Terhadap Karakteristik Fisik dan Mekanik. *Jurnal Teknosains Pangan* **2013**, *2*, 73–79.

48. Warkoyo, W.; Rahardjo, B.; Marseno, D.W.; Karyadi, J.N.W. Sifat fisik, mekanik dan barrier edible film berbasis pati umbi kimpul (Xanthosoma sagittifolium) yang diinkorporasi dengan kalium sorbat. *Agritech* **2014**, *34*, 72–81.

49. Pelissari, F.M.; Grossmann, M.V.; Yamashita, F.; Pineda, E.A.G. Antimicrobial, mechanical, and barrier properties of cassava starch-chitosan films incorporated with oregano essential oil. *J. Agric. Food Chem.* **2009**, *57*, 7499–7504. [CrossRef] [PubMed]

50. Song, X.; Cheng, L.; Tan, L. Edible iron yam and maize starch convenient food flavoring packaging films with lemon essential oil as plasticization. *Food Sci. Technol.* **2018**, *39*, 971–979. [CrossRef]

51. Péroval, C.; Debeaufort, F.; Despré, D.; Voilley, A. Edible arabinoxylan-based films. 1. Effects of lipid type on water vapor permeability, film structure, and other physical characteristics. *J. Agric. Food Chem.* **2002**, *50*, 3977–3983. [CrossRef]

52. Souza, A.C.; Goto, G.E.O.; Mainardi, J.A.; Coelho, A.C.V.; Tadini, C.C. Cassava starch composite films incorporated with cinnamon essential oil: Antimicrobial activity, microstructure, mechanical and barrier properties. *LWT Food Sci. Technol.* **2013**, *54*, 346–352. [CrossRef]

53. Galus, S. Functional properties of soy protein isolate edible films as affected by rapeseed oil concentration. *Food Hydrocoll.* **2018**, *85*, 233–241. [CrossRef]
54. Bertuzzi, M.A.; Vidaurre, E.C.; Armada, M.; Gottifredi, J.C. Water vapor permeability of edible starch based films. *J. Food Eng.* **2007**, *80*, 972–978. [CrossRef]
55. Shen, X.L.; Wu, J.M.; Chen, Y.; Zhao, G. Antimicrobial and physical properties of sweet potato starch films incorporated with potassium sorbate or chitosan. *Food Hydrocoll.* **2010**, *24*, 285–290. [CrossRef]
56. Sothornvit, R.; Pitak, N. Oxygen permeability and mechanical properties of banana films. *Food Res. Int.* **2007**, *40*, 365–370. [CrossRef]
57. Siracusa, V.; Romani, S.; Gigli, M.; Mannozzi, C.; Cecchini, J.; Tylewicz, U.; Lotti, N. Characterization of Active Edible Films based on Citral Essential Oil, Alginate and Pectin. *Materials* **2018**, *11*, 1980. [CrossRef] [PubMed]
58. Atarés, L.; Chiralt, A. Essential oils as additives in biodegradable films and coatings for active food packaging. *Trends Food Sci. Technol.* **2016**, *48*, 51–62. [CrossRef]
59. López, C.A.; de Vries, A.H.; Marrink, S.J. Amylose folding under the influence of lipids. *Carbohydr. Res.* **2012**, *364*, 1–7. [CrossRef] [PubMed]
60. Talón, E.; Vargas, M.; Chiralt, A.; González-Martínez, C. Antioxidant starch-based films with encapsulated eugenol. Application to sunflower oil preservation. *Food Sci. Technol.* **2019**, *113*, 108290. [CrossRef]
61. Shojaee-Aliabadi, S.; Hosseini, H.; Mohammadifar, M.A.; Mohammadi, A.; Ghasemlou, M.; Ojagh, S.M.; Khaksar, R. Characterization of antioxidant-antimicrobial κ-carrageenan films containing Satureja hortensis essential oil. *Int. J. Biol. Macromol.* **2013**, *52*, 116–124. [CrossRef] [PubMed]
62. Fernandes de Oliveira, A.; Sousa Pinheiro, L.; Souto Pereira, C.; Neves Matias, W. Albuquerque Gomes, R.; Souza Chaves, O.; Simões de Assis, T. Total phenolic content and antioxidant activity of some Malvaceae family species. *Antioxidants* **2012**, *1*, 33–43. [CrossRef] [PubMed]
63. Baser, K.H.C.; Buchbauer, G. *Handbook of Essential Oils: Science, Technology, and Applications*, 2nd ed.; CRC Press: Boca Raton, FL, USA, 2015; pp. 1–320.
64. Alma, M.H.; Ertas, M.; Nitz, S.; Kollmannsberger, H. Chemical composition and content of essential oil from the bud of cultivated Turkish clove (*Syzygium aromaticum* L.). *BioResources* **2007**, *2*, 265–269.
65. Jirovetz, L.; Buchbauer, G.; Stoilova, I.; Stoyanova, A.; Krastanov, A.; Schmidt, E. Chemical composition and antioxidant properties of clove leaf essential oil. *J. Agric. Food Chem.* **2006**, *54*, 6303–6307. [CrossRef]
66. Dashipour, A.; Khaksar, R.; Hosseini, H.; Shojaee-Aliabadi, S.; Kiandokht, G. Physical, antioxidant and antimicrobial characteristics of carboxymethyl cellulose edible film cooperated with clove essential oil. *Zahedan J. Res. Med. Sci.* **2014**, *16*, 34–42.
67. Burt, S.A.; Reinders, R.D. Antibacterial activity of selected plant essential oils against Escherichia coli O157: H7. *Lett. Appl. Microbiol.* **2003**, *36*, 162–167. [CrossRef]
68. Devi, K.P.; Nisha, S.A.; Sakthivel, R.; Pandian, S.K. Eugenol (an essential oil of clove) acts as an antibacterial agent against Salmonella typhi by disrupting the cellular membrane. *J. Ethnopharmacol.* **2010**, *130*, 107–115. [CrossRef] [PubMed]
69. Mahcene, Z.; Khelil, A.; Hasni, S.; Akman, P.K.; Bozkurt, F.; Birech, K.; Goudjil, M.B.; Tornuk, F. Development and characterization of sodium alginate based active edible films incorporated with essential oils of some medicinal plants. *Int. J. Biol. Macromol.* **2020**, *145*, 124–132. [CrossRef] [PubMed]
70. Atanasova-Pancevska, N.; Bogdanov, J.; Kungulovski, D. In Vitro Antimicrobial Activity and Chemical Composition of Two Essential Oils and Eugenol from Flower Buds of *Eugenia caryophyllata*. *Open Biol. Sci. J.* **2017**, *3*, 16–25. [CrossRef]
71. Mohamed, S.G.; Badri, A.M. Antimicrobial Activity of Syzygium aromaticum and Citrus aurantifolia Essential Oils against Some Microbes in Khartoum, Sudan. *EC Microbiol.* **2017**, *12*, 253–259.

 *foods*

Communication

HPTLC-DESI-HRMS-Based Profiling of Anthraquinones in Complex Mixtures—A Proof-of-Concept Study Using Crude Extracts of Chilean Mushrooms

Annegret Laub [1], Ann-Katrin Sendatzki [1], Götz Palfner [2], Ludger A. Wessjohann [1], Jürgen Schmidt [1] and Norbert Arnold [1,*]

[1] Department of Bioorganic Chemistry, Leibniz Institute of Plant Biochemistry, Weinberg 3, D-06120 Halle (Saale), Germany; alaub@ipb-halle.de (A.L.); asend@web.de (A.-K.S.); wessjohann@ipb-halle.de (L.A.W.); jschmidt@ipb-halle.de (J.S.)

[2] Departamento de Botanica, Facultad de Ciencias Naturales y Oceanograficas, Universidad de Concepcion, Casilla 160-C, Concepcion, Chile; gpalfner@udec.cl

* Correspondence: narnold@ipb-halle.de; Tel.: +49-(0)-345-5582-1310

Received: 20 January 2020; Accepted: 30 January 2020; Published: 6 February 2020

Abstract: High-performance thin-layer chromatography (HPTLC) coupled with negative ion desorption electrospray ionization high-resolution mass spectrometry (DESI-HRMS) was used for the analysis of anthraquinones in complex crude extracts of Chilean dermocyboid Cortinarii. For this proof-of-concept study, the known anthraquinones emodin, physcion, endocrocin, dermolutein, hypericin, and skyrin were identified by their elemental composition. HRMS also allowed the differentiation of the investigated anthraquinones from accompanying compounds with the same nominal mass in the crude extracts. An investigation of the characteristic fragmentation pattern of skyrin in comparison with a reference compound showed, exemplarily, the feasibility of the method for the determination of these coloring, bioactive and chemotaxonomically important marker compounds. Accordingly, we demonstrate that the coupling of HPTLC with DESI-HRMS represents an advanced and efficient technique for the detection of anthraquinones in complex matrices. This analytical approach may be applied in the field of anthraquinone-containing food and plants such as *Rheum* spp. (rhubarb), *Aloe* spp., *Morinda* spp., *Cassia* spp. and others. Furthermore, the described method can be suitable for the analysis of anthraquinone-based colorants and dyes, which are used in the food, cosmetic, and pharmaceutical industry.

Keywords: HPTLC-MS coupling; HPTLC; negative ion DESI-HR-MS/MS; anthraquinones; Chilean mushrooms; genus *Cortinarius*

1. Introduction

Anthraquinones represent a large family of naturally occurring pigments, which are produced by plants, microbes, lichens, insects, and fungi [1]. Besides their coloring properties, these natural products exhibit a broad range of bioactivities such as antibacterial, antiparasitic, anti-inflammatory, fungicidal, insecticidal, laxative, antiviral, and anticancer but also DNA intercalating properties [2–7]. The chemical structure of anthraquinones is based on an anthracene skeleton with two keto groups in position 9 and 10. The basic core unit can be further substituted at various positions and connected with sugar molecules, forming the corresponding glycosides [8,9].

In the literature, about 700 anthraquinone derivatives are described, in which emodin, physcion, catenarin, and rhein are the most frequently reported [9–11]. Two hundred of these are described for flowering plants, which also occur in edible plants and vegetables such as *Rheum*, *Aloe* and *Cassia* species, while the remaining ones are produced by lichens and fungi [7,8,12].

The genus *Cortinarius* (including Dermocybe) is one of the most diverse genera of basidiomycetous fungi containing a great variety of anthraquinones [13–15]. The occurrence and distribution of these pigments is closely linked to species diversity and allows their use as chemotaxonomic marker compounds in species delimination [16–22].

The analysis of anthraquinones is of interest due to their wide range of application. A continuous improvement of the analytical techniques is needed to overcome difficulties with respect to interference with various types of matrices and low abundance of the analytes within complex mixtures [7].

Thin-layer chromatography is an effective method for the chromatographic separation of anthraquinones [23–26]. Furthermore, several mass spectrometry-based methods have been developed for a deep analysis of anthraquinones providing characteristic $[M-H]^-$ ions in negative ion mode [27–29].

Desorption electrospray ionization mass spectrometry (DESI-MS) represents a powerful ambient ionization mass spectrometric technique, which enables a direct ionization of analytes from surfaces with subsequent mass spectrometric detection [30–32]. The coupling of DESI-MS with high-performance thin-layer chromatography (HPTLC) provides a robust methodological approach for the separation and highly sensitive detection of secondary metabolites in plants and fungi [33–35]. Furthermore, this method is suitable for the fingerprint analysis of crude extracts in natural product research [36,37]. Recently, the detection of excreted polyhydroxyanthraquinones from the surface of fungal culture agar plates using DESI-MS in negative ion mode was reported [38].

In the present paper, we report the development of a rapid profiling method of anthraquinones, exemplified with the analysis of different crude extracts from Chilean dermocyboid Cortinarii concerning their anthraquinone pattern based on the combination of HPTLC with negative ion DESI-HRMS. For this proof-of-concept study, extracts from fruiting bodies of six dermocyboid Cortinarii were investigated for the occurrence of the known anthraquinones emodin, physcion, endocrocin, dermolutein, hypericin, and skyrin. Furthermore, the possibility of performing MS/MS experiments on the desorbed analytes directly from the HPTLC plate was exemplarily shown for the bisanthraquinone skyrin in comparison with data obtained from direct-infusion MS experiments.

2. Materials and Methods

2.1. Reagents and Chemicals

The authentic reference compounds endocrocin (**3**), hypericin (**5**) and skyrin (**6**) were available from the in-house compound library of the Department of Bioorganic Chemistry, Leibniz Institute of Plant Biochemistry (IPB), Halle (Saale), Germany. Methanol and toluene were used at analytical grade. Ethyl formate was purchased from Merck (Darmstadt, Germany) and formic acid from Roth (Karlsruhe, Germany). LC-MS grade methanol was obtained from Merck (Darmstadt, Germany), and purified water was prepared by Merck Millipore Milli-Q equipment (Darmstadt, Germany).

2.2. Sampling Sites and Extraction

Fruiting bodies of *C.* (D.) *austronanceiensis*, *C.* (D.) *icterina*, *C.* (D.) *icterinula*, *C.* (D.) *obscuro-olivea*, *C.* (D.) spec., and *C.* (D.) *viridulifolius* were collected in Chile (detailed information see Table S1). Voucher specimens are deposited in the Fungarium of Concepción University (CONC-F). A duplicate is deposited at the Leibniz Institute of Plant Biochemistry.

Air-dried fruiting bodies (2 g) were homogenized using 15 mL of acetone in a blender followed by an ultrasonic extraction for 15 min to remove interfering compounds such as fatty acids from the material. After vacuum-supported filtration, the fungal material residue was further extracted twice with 15 mL methanol each. The resulting extracts were filtrated and dried under reduced pressure using a rotary evaporator. The crude methanolic extracts were redissolved in methanol and directly spotted on the HPTLC plate for chromatographic separation.

2.3. HPTLC

HPTLC was performed on Glass HPTLC Silica gel 60 F_{254} plates (10 × 10 cm, layer thickness 150–200 µm, Merck) using a mixture of toluene, ethyl formate, and formic acid (10:5:3; *v/v/v*) as a mobile phase. After air drying, the developed plates were parted by a glass cutter and subjected to the mass spectral analysis. For documentation and Rf-value determination, a CAMAG TLC visualizer (CAMAG, Muttenz, Switzerland) was used with the software winCATS (version 1.4.9.2001, CAMAG, Switzerland).

2.4. DESI-Orbitrap-MS and MS^2

All experiments were performed using a 2D-DESI source (Omnispray System OS-3201, Prosolia, Indianapolis, IN, USA) coupled to an Orbitrap Elite mass spectrometer (Thermo Fisher Scientific, Bremen, Germany) operated in the negative ion mode. The DESI source settings were as follows: spray voltage, 3 kV; solvent flow rate, 2 µL/min; nebulizing gas (nitrogen), pressure, 7 bar; tip-to-surface distance, 2–2.5 mm; tip-to-inlet distance, 3.5 mm; incident angle (relative to the surface plane), 55°. The DESI spray solvent was 50:50 (*v/v*) methanol/water. MS experiments were performed by continuously scanning every HPTLC band in the y-direction (from Rf 0 to 1.0) at a surface velocity of 200 µm/s while acquiring mass spectra in full scan mode (*m/z* 150–1500; resolution 30,000) and 150 µm/s in MS^2 mode. Collision-induced dissociation was performed using normalized collision energies (NCE) of 35 and 50 (arbitrary units) and an isolation width of ± 2 Da. The data were evaluated using the software Xcalibur 2.2 SP1 (Thermo Fisher Scientific).

3. Results and Discussion

3.1. Method Development

The normal-phase HPTLC plates were developed for 55 mm in one dimension to enable the separation of anthraquinones according to their polarity (Figure 1). The geometry of the source, the composition of the spray solvent, the flow rate as well as the scanning rate were optimized for the analysis. To enhance the ionization and desorption efficiency, different mixtures of methanol and water (with and without formic acid as additive) were tested as spray solvents. A mixture of methanol and water of 1:1 (*v/v*) yielded the best results. During optimization, a flow rate of 2 µL/min showed good results to obtain adequate signal intensities. On the other hand, higher flow rates led to a partial detachment of silica gel particles. Additionally, different velocities of the DESI spray head were tested to obtain an efficient number of precursor ions for the MS^2 experiments. Lower scan rates led to better signal intensities due to the better desorption of the analytes from the surface of the HPTLC plates. Therefore, we used a lower velocity for the MS^2 experiments in the final measurements than in the full scan runs. Each band was recorded by scanning the surface in the y-direction (Rf 0 to 1.0) with an automated DESI source coupled to an Orbitrap Elite mass spectrometer within a total run time of 4.6 min. Before starting the experiment, the spray head was positioned on the application line of the HPTLC followed by the manual start of the MS measurement.

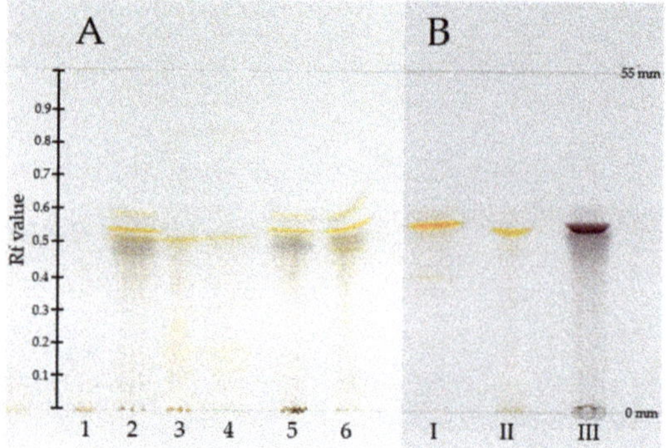

Figure 1. High-performance thin-layer chromatograph (HPTLC) of (**A**) methanolic crude extracts of *C.* (D.) spec. (**1**), *C.* (D.) *austronanceiensis* (**2**), *C.* (D.) *icterina* (**3**), *C.* (D.) *icterinula* (**4**), *C. obscuro-olivea* (**5**), *C. viridulifolius* (**6**) and (**B**) reference compound skyrin (**6**, I), endocrocin (**3**, II), and hypericin (**5**, III) (mobile phase: toluene, ethyl formate, and formic acid (10:5:3; *v/v/v*) distance from application line to solvent front: 55 mm).

3.2. Profiling of Anthraquinones in Crude Extracts

The pigment pattern of the methanolic extracts of dermocyboid Cortinarii *Cortinarius* (Dermocybe) *austronanceiensis*, *C.* (D.) *icterina*, *C.* (D.) *icterinula*, *C.* (D.) *obscuro-olivea*, *C.* (D.) spec., and *C.* (D.) *viridulifolius* (Figure 1) was analyzed by high-performance thin-layer chromatography (HPTLC) coupled to desorption electrospray ionization (DESI) mass spectrometry in the negative ion mode. An unspotted HPTLC band was scanned to assign background related peaks (Figure S1) and to ensure the absence of the target compounds before applying the crude extracts on the HPTLC plate. No anthraquinone-related peaks could be detected by scanning the empty band on the HPTLC plate after running with the solvent system. This is demonstrated by the extracted ion chromatograms based on the theoretical calculated *m/z* value of the [M-H]$^-$ ions using an 25 ppm window (four decimals) (Figure S2). The established analytical method was applied to identify anthraquinones 1–6 (Figure 2). These anthraquinones were chosen for this proof-of-concept study because their occurrence in different *Cortinarius* and *Dermoybe* species is described in the literature [15]. The assignment of the structures is based on their elemental composition determined by high-resolution mass spectrometry (HRMS) (Table 1 and Table S2).

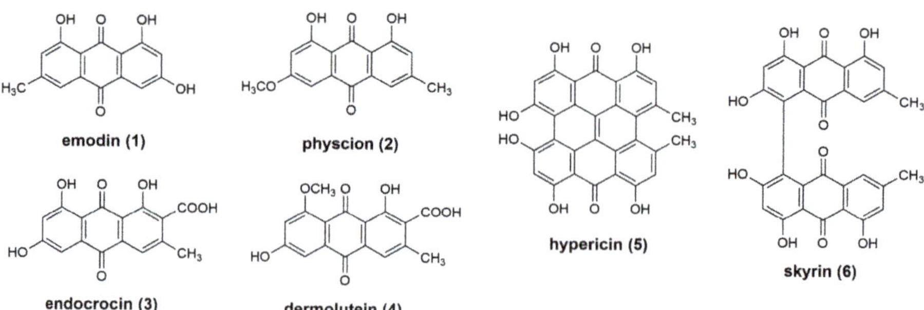

Figure 2. Structures of investigated anthraquinones **1–6**.

Table 1. Detected anthraquinones (**1–6**) using HPTLC-desorption electrospray ionization (DESI)-high-resolution mass spectroscopy (HRMS).

No.	Elemental Composition $[M\text{-}H]^-$	Theoretical m/z $[M\text{-}H]^-$	C. (D.) *austronanceiensis*	C. (D.) *icterina*	C. (D.) *icterinula*	C. (D.) *obscuro-olivea*	C. (D.) *spec.*	C. (D.) *viridulifolius*
1	$C_{15}H_9O_5^-$	269.0455	+	+	+	+	+	+
2	$C_{16}H_{11}O_5^-$	283.0612	+	+	n.d.	+	+	+
3	$C_{16}H_9O_7^-$	313.0354	+	+	+	+	n.d.	+
4	$C_{17}H_{11}O_7^-$	327.0510	+	+	+	+	+	+
5	$C_{30}H_{15}O_8^-$	503.0772	+	n.d.	n.d.	+	n.d.	+
6	$C_{30}H_{17}O_{10}^-$	537.0827	+	n.d.	n.d.	+	n.d.	+

[1] n.d. = not detected; + = detected.

The negative ion DESI mass spectra of the methanolic extract of *C.* (D.) *austronanceiensis* afforded characteristic deprotonated ions of emodin (1, $[M\text{-}H]^-$ at m/z 269.0450 calcd for $C_{15}H_9O_5^-$ 269.0455), physcion (2, $[M\text{-}H]^-$ at m/z 283.0611 calcd for $C_{16}H_{11}O_5^-$ 283.0612), endocrocin (3, $[M\text{-}H]^-$ at m/z 313.0349 calcd for $C_{16}H_9O_7^-$ 313.0354), dermolutein (4, $[M\text{-}H]^-$ at m/z 327.0505 calcd for $C_{17}H_{11}O_7^-$ 327.0510), hypericin (5, $[M\text{-}H]^-$ at m/z 503.0763 calcd for $C_{30}H_{15}O_8^-$ 503.0772) and skyrin (6, $[M\text{-}H]^-$ at m/z 537.0817 calcd for $C_{30}H_{17}O_{10}^-$ 537.0827) (Figure 3A). For the data evaluation, the target m/z values were extracted from the total ion chromatogram using a 25 ppm mass window with a mass accuracy of four decimals to obtain the corresponding extracted ion chromatograms (EICs) for each analyte. The EICs for the methanolic extracts of *C.* (D.) *icterina*, *C.* (D.) *icterinula*, *C.* (D.) *obscuro-olivea*, *C.* (D.) *spec.*, and *C.* (D.) *viridulifolius* are shown in Figures S3–S7, and the presences of the different target analytes within the extracts are visualized in Table 1. Due to the resolving power of the orbitrap detector, a differentiation of isobaric ions was possible as shown in the EIC of dermolutein (4, Figure 3B). The anthraquinone peak m/z 327.0505 is clearly separated from other accompanying ions at the same nominal mass using a resolution of 30,000.

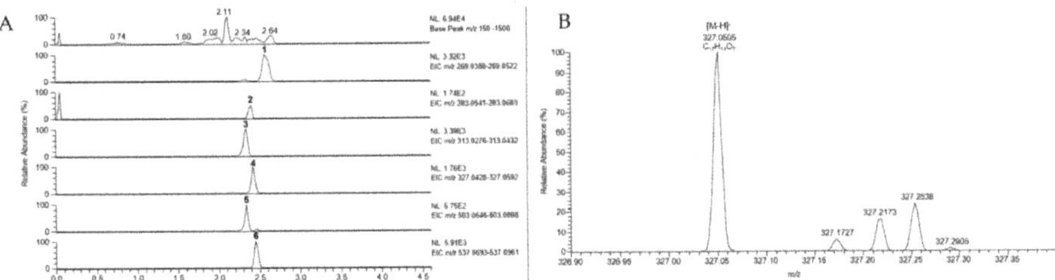

Figure 3. (**A**) Extracted ion chromatograms (EIC, mass window: 25 ppm) of anthraquinones 1-6 from crude extract of *Cortinarius* (D.) *austronanceiensis*, (**B**) Extracted ion chromatogram (EIC) of dermolutein (3, m/z 327) acquired during DESI-HR-MS measurement of methanolic extract from *C.* (D.) *austronanceiensis*.

Comparing the pigment patterns of the different fungal extracts (Table 1), all targets could be detected in the methanolic extracts of *Cortinarius* (D.) *austronanceiensis*, *C.* (D.) *obscuro-olivea* and *C.* (D.) *viridulifolius*. The naphthodianthrone hypericin (5) and the bisanthraquinone skyrin (6) were not detectable along the HPTLC bands of *C.* (D.) *icterina*, *C.* (D.) *icterinula* and *C.* (D.) spec.

Based on the retention time and the velocity of scanning the HPTLC bands (see Equation (1)), Rf values can be calculated and compared with the Rf values determined directly from the HPTLC plate (Table 2). The results of the developed HPTLC plates of the extracts and the reference compounds (see Figures S7–S10) were reproducible and comparable, exemplified based on the extracted ion chromatograms of endocrocin (Figure S11). Therefore, the determination of Rf values based on the retention time of the HPTLC-DESI-MS measurements of UV/VIS inactive analytes could be possible.

$$\text{Rf} = t_R \text{ (min)} \times \text{velocity (mm/s)} \times 60 \times 1/\text{distance from application line to solvent front (mm)}$$
$$\text{Rf} = t_R \times 0.200 \text{ mm/s} \times 60 \times 1/55 \text{ mm} \tag{1}$$

Table 2. Rf value and calculation from crude extract of *C. (D.) austronanciensis* (Figure 1, band 2).

Compound	Rf (experimental)	t_R DESI (min)	Rf (calculated)	Spot Color Visible Light	Spot Color UV Light (254 nm)	Spot Color UV Light (366 nm)
1	0.58	2.57	0.56	yellow	dark	orange
2	0.54	2.40	0.52	yellow	dark	orange
3 *	0.5	2.34	0.51	yellow	dark	orange
4	0.49	2.42	0.53	yellow	dark	red
5 *	0.5	2.34	0.51	black	dark	red
6 *	0.53	2.45	0.53	yellow-orange	dark	red brown

* Confirmed with reference compound.

3.3. Structural Characterization Using MS^2 Experiments

As an example, the fragmentation behavior of the bisanthraquinone skyrin (6) was investigated by a MS^2 measurement compared with the results obtained directly from the extract, data of a reference compound measured by HPTLC-DESI-HRMS and with direct infusion DESI-HRMS (Figure 4A–C, Table S3). Skyrin (6), in its MS^2 spectrum, shows a base peak ion at m/z 493.0923 ([M-H-CO$_2$]$^-$, calcd for C$_{29}$H$_{17}$O$_8^-$ 493.0929, Figure 4A, Table S3). Furthermore, a loss of carbon suboxide (C$_3$O$_2$) is observed at m/z 469.0926 (calcd for C$_{27}$H$_{17}$O$_8^-$ 469.0929), indicating a 1,3-dihydroxybenzene feature, which is also described for flavones and other polyphenols [39,40]. The obtained data are in good agreement with the reported MS^2 data of skyrin [41].

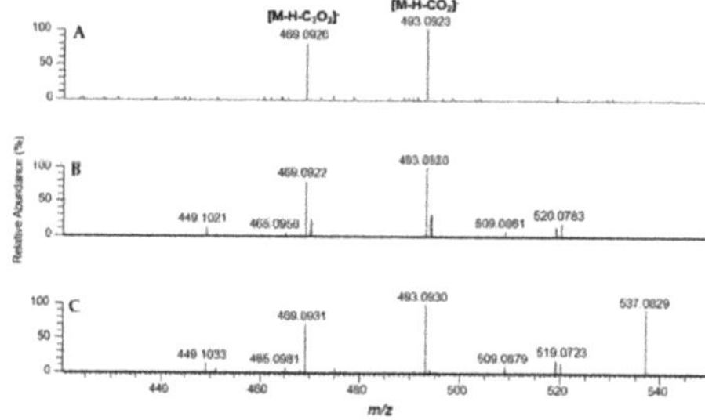

Figure 4. HPTLC-DESI-HR-MS^2 data of skyrin (6); (**A**) from fungal extract of *Cortinarius* (D.) *austronanceiensis* (NCE: 50%); (**B**) skyrin standard (NCE 35%); (**C**) direct infusion ESI-HR-MS^2 (NCE 30%).

4. Conclusions

Crude extracts of six Chilean dermocyboid Cortinarii were investigated by HPTLC-negative ion DESI-HRMS concerning the occurrence of the anthraquinones physcion (1), emodin (2), endocrocin (3), dermolutein (4), hypericin (5), and skyrin (6). The compounds were identified by their elemental composition. It should be pointed out that the high-resolution mass spectrometry (HRMS) approach also allows a mass spectral distinction of isobaric ions as demonstrated for the detection of dermolutein (4) whose nominal mass is accompanied by other compounds in the crude extract. Furthermore, the implementation of fragmentation experiments (MS^2) for anthraquinones on HPTLC surfaces is possible, as exemplarily shown for the detection of skyrin (6) in the extract of *C. (D) austronanceiensis*, and could be a valuable tool for the presence of these compound classes. The corresponding results are in good agreement with the data obtained by direct infusion and in comparison with the LC-MS data reported in literature.

HPTLC provides good separation efficiencies and can be performed in an automated and controlled way with respect to the sample application and the development of the plate. In classical approaches, a derivatization of the HPTLC plate is needed; however, combined with DESI-MS, this step is not required. Although the separation power of HPTLC is lower than in (U)HPLC, several analyses can performed with one plate and within a short analysis time. In case of the presented approach a HPTLC plate (total length 100 mm) with a developing length of 55 mm, and a total scanning time of only 4.6 min was necessary to obtain the presented results. After the extraction of the material, no further sample preparation steps are necessary, and the crude extracts can be directly applied to the plate, representing an advantage compared with other analytical techniques.

In summary, the obtained results illustrate the feasibility and capacity of HPTLC-DESI-HRMS to provide a rapid first screening method for the analysis of anthraquinones in complex mixtures, which may be used in the analysis of anthraquinones in food, plants, fungi, dyes, and cosmetic and pharmaceutical products.

Supplementary Materials: The following are available online at http://www.mdpi.com/2304-8158/9/2/156/s1, Figure S1: (A) Total ion chromatogram of an empty HPTLC band after development with an eluent system (toluene, ethyl formate, and formic acid (10:5:3; *v/v/v*)), (B) Corresponding full MS spectrum to A (averaged over Rt 0–4.6 min) showing background related peaks. Figure S2: Total ion (A), base peak (B) and extracted ion chromatograms of an unspotted HPTLC band showing no anthraquinone related peaks (C–H)., Figure S3: Base peak chromatogram (A) and extracted ion chromatograms (EICs, B–G) based on the theoretical masses of the investigated anthraquinones (1–6) obtained from the methanolic crude extract of *Cortinarius* (D.) *icterina.*, Figure S4: Base peak chromatogram (A) and extracted ion chromatograms (EICs, B–G) based on the theoretical masses of the investigated anthraquinones (1–6) obtained from the methanolic crude extract of *Cortinarius* (D.) *icterinula.*, Figure S5: Base peak chromatogram (A) and extracted ion chromatograms (EICs, B–G) based on the theoretical masses of the investigated anthraquinones (1–6) obtained from the methanolic crude extract of *Cortinarius* (D.) *obscuro-olivea.*, Figure S6: Base peak chromatogram (A) and extracted ion chromatograms (EICs, B–G) based on the theoretical masses of the investigated anthraquinones (1–6) obtained from the methanolic crude extract of *Cortinarius* (D.) spec., Figure S7: Base peak chromatogram (A) and extracted ion chromatograms (EICs, B–G) based on the theoretical masses of the investigated anthraquinones (1–6) obtained from the methanolic crude extract of of *Cortinarius* (D.) *viridulifolius.*, Figure S8: Total ion and extracted ion chromatogram (EIC) of reference compound skyrin (6) and the corresponding HRMS spectrum., Figure S9: Total ion and extracted ion chromatogram EIC) of reference compound endocrocin (3) and the corresponding HRMS spectrum., Figure S10: Total ion and extracted ion chromatogram (EIC) of reference compound hypericin (5) and the corresponding HRMS spectrum., Figure S11: Comparison of extracted ion chromatograms (EICs) of endocrocin (3, *m/z* 313) acquired during DESI-HR-MS measurement of methanolic extract from *C.* (D.) species and the reference compound., Table S1: Origin of fungal material. Table S2: Detected anthraquinones (1–6), their elemental composition and exact masses., Table S3: Key ions in the negative ion ESI-MS^n spectra of skyrin (6). The spectroscopic data are available at RADAR [42].

Author Contributions: Conceptualization, A.L., J.S. and N.A.; methodology, A.L., A.-K.S.; validation, A.L., J.S.; formal analysis, A.L., A.-K.S.; investigation, A.L., A.-K.S.; resources, G.P., N.A.; data curation, A.L.; writing—original draft preparation, A.L., J.S., N.A.; writing—review and editing, J.S., N.A., L.A.W.; supervision, J.S., N.A., L.A.W.; project administration, G.P., N.A.; funding acquisition, G.P., N.A. All authors have read and agreed to the published version of the manuscript.

Funding: This research was funded by BMBF, grant number 01DN12107, and CONICYT, grant number PCI 2011-609.

Conflicts of Interest: The authors declare no conflict of interest.

References

1. Thomson, R.H. *Naturally Occurring Quinones IV*, 4th ed.; Springer: Dordrecht, The Netherlands, 1997; pp. 309–483.
2. Wakulinski, W.; Kachlicki, P.; Sobiczewski, P.; Schollenberger, M.; Zamorski, C.; Lotocka, B.; Sarova, J. Catenarin production by isolates of *Pyrenophora tritici-repentis* (Died.) Drechsler and its antimicrobial activity. *J. Phytopathol.* **2003**, *151*, 74–79. [CrossRef]
3. Reynolds, T. *Aloes—The Genus Aloe*, 1st ed.; CRC Press: Boca Raton, FL, USA, 2004.
4. Srinivas, G.; Babykutty, S.; Sathiadevan, P.P.; Srinivas, P. Molecular mechanism of emodin action: Transition from laxative ingredient to an antitumor agent. *Med. Res. Rev.* **2007**, *27*, 591–608. [CrossRef] [PubMed]
5. Locatelli, M. Anthraquinones: Analytical techniques as a novel tool to investigate on the triggering of biological targets. *Curr. Drug Targets* **2011**, *12*, 366–380. [CrossRef] [PubMed]
6. Chien, S.C.; Wu, Y.C.; Chen, Z.W.; Yang, W.C. Naturally occurring anthraquinones: Chemistry and therapeutic potential in autoimmune diabetes. *Evid. Based Complement. Alternat. Med.* **2015**, *2015*, 357357. [CrossRef]
7. Duval, J.; Pecher, V.; Poujol, M.; Lesellier, E. Research advances for the extraction, analysis and uses of anthraquinones: A review. *Ind. Crops Prod.* **2016**, *94*, 812–833. [CrossRef]
8. Seigler, D.S. *Plant Secondary Metabolism*; Springer: New York, NY, USA, 1998; p. 85.

9. Gessler, N.N.; Egorova, A.S.; Belozerskaya, T.A. Fungal anthraquinones. *Appl. Biochem. Microbiol.* **2013**, *49*, 85–99. [CrossRef]

10. Caro, Y.; Anamale, L.; Fouillaud, M.; Laurent, P.; Petit, T.; Dufosse, L. Natural hydroxyanthraquinoid pigments as potent food grade colorants: An overview. *Nat. Prod. Bioprospect.* **2012**, *2*, 174–193. [CrossRef]

11. Fouillaud, M.; Venkatachalam, M.; Girard-Valenciennes, E.; Caro, Y.; Dufosse, L. Anthraquinones and derivatives from marine-derived fungi: Structural diversity and selected biological activities. *Mar. Drugs* **2016**, *14*, 64. [CrossRef]

12. Fouillaud, M.; Caro, Y.; Venkatachalam, M.; Grondin, I.; Laurent, D. Anthraquinones. In *Phenolic Compounds in Food—Characterization and Analysis*; Nollet, L.M.L., Gutierrez-Uribe, J.A., Eds.; CRC Press: Boca Raton, FL, USA, 2018; pp. 130–170.

13. Gill, M.; Steglich, W. Pigments of fungi (macromycetes). *Progr. Chem. Org. Chem. Nat. Prod.* **1987**, *51*, 1–317.

14. Gill, M. New pigments of *Cortinarius* Fr. And *Dermocybe* (Fr.) Wünsche (Agaricales) from Australia and New Zealand. *Beih. Sydowia* **1995**, *10*, 73–87.

15. Keller, G.; Moser, M.; Horak, E.; Steglich, W. Chemotaxonomic investigations of species of *Dermocybe* (Fr. Wünsche (Agaricales) from New zealand, Papua New Guinea and Argentina. *Beih. Sydowia* **1988**, *10*, 101–126.

16. Steglich, W.; Austel, V. Die Struktur des Dermocybins und des Dermoglaucins. *Tetrahedron Lett.* **1966**, *26*, 3077–3079. [CrossRef]

17. Gruber, I. Anthrachinonfarbstoffe in der Gattung *Dermocybe* und Versuch ihrer Auswertung für die Systematik. *Zeitschr. Pilzk.* **1970**, *36*, 95–112.

18. Arnold, N.; Besl, A.; Bresinsky, A.; Kemmer, H. Notizen zur Chemotaxonomie der Gattung *Dermocybe* (Agaricales) und zu ihrem Vorkommen in Bayern. *Z. Mykol.* **1987**, *53*, 187–194.

19. Arnold, N. *Morphologisch-Anatomische und Chemische Untersuchungen an der Untergattung Telamonia (Cortinarius, Agaricales)*; IHW-Verlag: Eching, München, Germany, 1993.

20. Jones, R.H.; May, T.W. Pigment chemistry and morphology support recognition of *Cortinarius austrocinnabarinus* sp. nov. (fungi: Cortinariaceae) from Australia. *Muelleria* **2008**, *26*, 77–87.

21. Stefani, F.O.; Jones, R.H.; May, T.W. Concordance of seven gene genealogies compared to phenotypic data reveals multiple cryptic species in Australian dermocyboid *Cortinarius* (Agaricales). *Mol. Phylogenet. Evol.* **2014**, *71*, 249–260. [CrossRef] [PubMed]

22. Greff, A.; Porzel, A.; Schmidt, J.; Palfner, G.; Arnold, N. Pigment pattern of the chilean mushroom *Dermocybe nahuelbutensis* Garrido & E. Horak. *Rec. Nat. Prod.* **2017**, *11*, 547–551.

23. Shibata, S.; Takito, M.; Tanaka, O. Paper chromatography of anthraquinone pigments. *J. Am. Chem. Soc.* **1950**, *72*, 2789–2790. [CrossRef]

24. Kidd, C.B.M.; Caddy, B.; Robertson, J.; Tebbett, I.R. Thin-layer chromatography as an aid for identification of *Dermocybe* species of *Cortinarius*. *Trans. Br. Mycol. Soc.* **1985**, *85*, 213–221. [CrossRef]

25. Ma, X.; Chen, Y.; Hui, R. Analysis of anthraquinones in *Rheum franzenbachii* Münt (rhubarb) by thin-layer chromatography. *Chromatographia* **1989**, *27*, 465–466. [CrossRef]

26. Räisänen, R.; Björk, H.; Hynninen, H. Two-dimensional TLC separation and mass spectrometric identification of anthraquinones isolated from the fungus *Dermocybe sanguinea*. *Z. Naturforsch.* **2000**, *55c*, 195–202. [CrossRef] [PubMed]

27. Lin, C.-C.; Wu, C.-I.; Lin, T.-C.; Sheu, S.-J. Determination of 19 rhubarb constituents by high-performance liquid chromatography–ultraviolet–mass spectrometry. *J. Sep. Sci.* **2006**, *29*, 2584–2593. [CrossRef] [PubMed]

28. Ye, M.; Han, J.; Chen, H.; Zheng, J.; Guo, D. Analysis of phenolic compounds in rhubarbs using liquid chromatography coupled with electrospray ionization mass spectrometry. *J. Am. Soc. Mass. Spectrom.* **2007**, *18*, 82–91. [CrossRef] [PubMed]

29. Derksen, G.C.H.; Niederländer, H.A.G.; Van Beek, T.A. Analysis of anthraquinones in *Rubia tinctorium* L. by liquid chromatography coupled with diode-array UV and mass spectrometric detection. *J. Chromatogr. A* **2002**, *978*, 119–127. [CrossRef]

30. Takats, Z.; Wiseman, J.M.; Gologan, B.; Cooks, R.G. Mass spectrometry sampling under ambient conditions with desorption electrospray ionization. *Science* **2004**, *306*, 471–473. [CrossRef]

31. Takats, Z.; Wiseman, J.M.; Cooks, R.G. Ambient mass spectrometry using desorption electrospray ionization (DESI): Instrumentation, mechanisms and applications in forensics, chemistry, and biology. *J. Mass Spectrom.* **2005**, *40*, 1261–1275. [CrossRef]

32. Cooks, R.G.; Ouyang, Z.; Takats, Z.; Wiseman, J.M. Detection technologies. Ambient mass spectrometry. *Science* **2006**, *311*, 1566–1570. [CrossRef]

33. Van Berkel, G.J.; Tomkins, B.A.; Kertesz, V. Thin-layer chromatography/desorption electrospray ionization mass spectrometry: Investigation of goldenseal alkaloids. *Anal. Chem.* **2007**, *79*, 2778–2789. [CrossRef]

34. Lane, A.L.; Nyadong, L.; Galhena, A.S.; Shearer, T.L.; Stout, E.P.; Parry, R.M.; Kwasnik, M.; Wang, M.D.; Fernandez, F.M.; Kubanek, J. Desorption electrospray ionization mass spectrometry reveals surface-mediated antifungal chemical defense of a tropical seaweed. *PNAS* **2009**, *106*, 7314–7319. [CrossRef]

35. Nyadong, L.; Hohenstein, E.G.; Galhena, A.; Lane, A.L.; Kubanek, J.; Sherrill, C.D.; Fernandez, F.M. Reactive desorption electrospray ionization mass spectrometry (DESI-MS) of natural products of a marine alga. *Anal. Bioanal. Chem.* **2009**, *394*, 245–254. [CrossRef]

36. Bagatela, B.S.; Lopes, A.P.; Cabral, E.C.; Perazzo, F.F.; Ifa, D.R. High-performance thin-layer chromatography/desorption electrospray ionization mass spectrometry imaging of the crude extract from the peels of *Citrus aurantium* L. (Rutaceae). *Rapid Commun. Mass Spectrom.* **2015**, *29*, 1530–1534. [CrossRef] [PubMed]

37. Kennedy, J.H.; Wiseman, J.M. Direct analysis of *Salvia divinorum* leaves for salvinorin a by thin layer chromatography and desorption electrospray ionization multi-stage tandem mass spectrometry. *Rapid Commun. Mass Spectrom.* **2010**, *24*, 1305–1311. [CrossRef]

38. Figueroa, M.; Jarmusch, A.K.; Raja, H.A.; El-Elimat, T.; Kavanaugh, J.S.; Horswill, A.R.; Cooks, R.G.; Cech, N.B.; Oberlies, N.H. Polyhydroxyanthraquinones as quorum sensing inhibitors from the guttates of *Penicillium restrictum* and their analysis by desorption electrospray ionization mass spectrometry. *J. Nat. Prod.* **2014**, *77*, 1351–1358. [CrossRef] [PubMed]

39. Fabre, N.; Rustan, I.; De Hoffmann, E.; Quetin-Leclercq, J. Determination of flavone, flavonol, and flavanone aglycones by negative ion liquid chromatography electrospray ion trap mass spectrometry. *J. Am. Soc. Mass Spectrom.* **2001**, *12*, 707–715. [CrossRef]

40. Schmidt, J. Negative ion electrospray high-resolution tandem mass spectrometry of polyphenols. *J. Mass Spectrom.* **2016**, *51*, 33–43. [CrossRef]

41. Jahn, L.; Schafhauser, T.; Wibberg, D.; Ruckert, C.; Winkler, A.; Kulik, A.; Weber, T.; Flor, L.; van Pee, K.H.; Kalinowski, J.; et al. Linking secondary metabolites to biosynthesis genes in the fungal endophyte *Cyanodermella asteris*: The anti-cancer bisanthraquinone skyrin. *J. Biotechnol.* **2017**, *257*, 233–239. [CrossRef] [PubMed]

42. Laub, A.; Sendatzki, A.-K.; Schmidt, J.; Arnold, N. *Dataset: DESI-MS Data for "HPTLC-DESI-HRMS Based Profiling of Anthraquinones in Complex Mixtures—A Proof-Of-Concept Study using Crude Extracts of Chilean Mushrooms"*; RADAR (Reasearch Data Repository) v1.33; RADAR: Karlsruhe, Germany, 2015. [CrossRef]

MDPI

St. Alban-Anlage 66

4052 Basel

Switzerland

Tel. +41 61 683 77 34

Fax +41 61 302 89 18

www.mdpi.com

Foods Editorial Office

E-mail: foods@mdpi.com

www.mdpi.com/journal/foods